PROSTATE CANCER
MEETS
THE PROTON BEAM

PROSTATE CANCER MEETS THE PROTON BEAM

~~~

A Patient's Experience

Fuller Jones

Revised Edition

Foreword by Robert J. Marckini

ISBN 978-0-615-19243-7

Revised Edition
July 2008

Printed in the United States of America
On Acid-Free Paper

To Bob Marckini

and

The Brotherhood of the Balloon

Contents

Picture this: You are diagnosed with cancer.

Your doctor encourages major surgery and announces the possibility of blood loss, infection, and the high probability of impotence and incontinence.

Next, imagine you discover a treatment that involves no invasion of your body, no blood loss, no infection, and significantly fewer side effects, if any.

You would feel the weight of the world lifted from your shoulders.

That is what happens with proton patients.

~

Robert J. Marckini

Foreword

CANCER IS A DREADED DISEASE, but if you're going to have cancer, prostate cancer is the one to have. And if you catch it early, the whole ordeal will likely be just a bump in the road, especially if you choose the right treatment option.

If you are diagnosed with prostate cancer, the most important thing is to do your homework. Don't just accept the surgical option because your urologist—probably a surgeon—tells you he thinks it's best for you. There are numerous viable, non-surgical options with success rates as good as or better than surgery, but with fewer debilitating side effects. And the best treatment option for you may be one that your urologist has never heard of. This option is proton beam therapy, an FDA approved, Medicare reimbursed, time-tested treatment modality that is gathering attention and gaining momentum in the medical community.

My name is Bob Marckini, and my exhaustive research led me to this wonderful treatment modality in the year 2000. My experience with proton beam therapy prompted me to form a prostate cancer support group that has become an international network of thousands of prostate cancer patients from twenty-three countries. These men have joined together to promote proton beam therapy. Fuller Jones became one of these men in late 2006, after finding my Web site, and contacting me with questions about proton beam therapy.

As a retired NASA engineer, Fuller Jones had little trouble understanding the benefits of proton technology and then choosing it for his own prostate cancer treatment. During his treatment, he decided to promote proton therapy by writing a book about the healing power of protons.

Fuller's book, *Prostate Cancer Meets the Proton Beam,* is about his personal journey with prostate cancer. It details his evaluation of the treatment options, his choice of proton beam therapy, and his treatment experience at Loma Linda University Medical Center.

Foreword

The book is rich with valuable information about how to become an informed patient, including a synopsis of each of the major treatment alternatives, and a detailed description of proton beam therapy.

A comprehensive bibliography is also provided, with a treasure trove of books, journals and websites. These provide the reader with the means to verify the author's source information and to further research prostate cancer and the various options for treatment.

Finally, Fuller's book summarizes the history of the extraordinary institution that pioneered proton therapy, Loma Linda University Medical Center (LLUMC), in Southern California. This teaching hospital is at the forefront of several leading edge medical technologies, and has set a new standard for patient care. Twenty-five years ago LLUMC and the visionary leaders of the institution took great risk and gambled their future on proton beam technology. Today, the result is a remarkable facility that is being replicated all over the world.

The seminal work of these dedicated professionals proved their hypothesis that proton particles could be harnessed in a hospital setting, accelerated to nearly the speed of light, and delivered to a target with pinpoint accuracy, effectively destroying cancers of all kinds.

All this with no pain or discomfort, while leaving the patient with a high quality of life after the treatment ends. The Dr. James M. Slater Proton Treatment Center at Loma Linda has not only legitimized proton beam therapy, it has shown it to be the new platinum standard for treating prostate cancer.

Prostate Cancer Meets The Proton Beam is a nuts-and-bolts journal for the prostate cancer victim, as well as his spouse and family members. If you are a recently diagnosed patient, it will answer most of your questions and leave you with the sense that this dreaded disease can be conquered, and with protons, the knowledge that your quality of life need not be degraded.

ROBERT J. MARCKINI
Author of *You Can Beat Prostate Cancer*

Preface

THE PRIMARY PURPOSE of this book is to awaken you to the possibility that you, or someone you know or love may develop some form of cancer, and if you are a man, that prostate cancer is a very real and present danger.

The secondary purpose of this book is to make known the fact that there is a technologically advanced cancer treatment that is presently not well known, and also is not widely recommended by the medical community. This is true even though it has been used in the United States for prostate cancer since 1990 with a success rate comparable to all other options, including radical prostatectomy. It is most definitely a treatment alternative that should be considered because, compared to the other treatment modalities, the side effects are generally minimal.

This treatment, *proton beam therapy*, which is truly state-of-the-art radiation medicine, is based on the elegant physics of the proton. This treatment may be beneficial not only for prostate cancer, but for many other forms of the disease so long as metastasis has not occurred.

In this book, much of what I learned while trying to decide on a form of prostate cancer treatment is provided, including a discussion of the various generally accepted methods. My personal viewpoint of the decision process is summarized, along with my own experiences while receiving proton therapy.

A summary of the "Legacy" of Loma Linda University Medical Center and other facts are also included. In addition, I try to provide some of the background and history of proton therapy and to explain in layman's terms some of the technical aspects of the very complex systems necessary to administer the treatments.

This book is intended to provide assistance to any person concerned about or diagnosed with cancer, in learning about proton beam therapy as one possible treatment modality.

The author is a retired engineer, not a medical doctor, and the information, comments, and/or advice in this book represent the author's personal experiences and opinions only. Any decision to utilize proton beam therapy for treatment of cancer or any other medical condition should be made in conjunction with qualified professional physicians and/or radiation oncologists.

Please recognize that this book is a *reference* only, to be used by individuals while searching for and deciding on a treatment for his or her own particular condition and situation. A patient should use all the research facts that he has found along with his best judgment on how the treatment selected will likely affect his own priorities and especially his quality of life, in making treatment decisions.

All factors involved in the decision process should be discussed not only with the patient's medical doctors, oncologists, or other professional medical specialists, but with close family members, loved ones, or friends as well.

Becoming an Informed Patient

THE SINGLE MOST IMPORTANT THING that one must understand about cancer is that early detection is the key to survival. If detected early many forms of cancer are treatable. Many times, a cure, or stopping (or at least slowing) the progress of the disease is possible.

The earlier that cancer of any form is detected, the better the chances for successful treatment and survival.

While the main focus of this book is prostate cancer, some additional facts about cancer that are important for the reader to recognize are as follows:

- Over 500,000 people die of cancer in the United States every year; that is more than 1,300 each day.

- Cancer causes more deaths than anything else except heart disease.

- Lung cancer is the leading cause of cancer-related death in the United States for both men and women.

- Prostate cancer is the second leading cause of cancer-related death in men.

- Breast cancer is the second leading cause of cancer-related death in women.

- Colorectal cancers are the third most frequent cause of death in both men and women.

- Cancer is the second leading cause of death in children, even though it is relatively rare. Accidents are the first.

There are many forms of cancer, and the disease can affect almost every part of the human body. It spares no one, regardless of age, sex, or race.

Prostate cancer is the primary topic of this book. This is because I am a prostate cancer patient and survivor, and I feel that it is my duty to try to alert those men who are not aware of their odds of contracting the disease.

What is prostate cancer? It is a malignant invasion of the cells in the prostate gland, occurring when normal prostate cells mutate and begin to multiply. The prostate (not prostrate) is a gland that is located at the lower end of the bladder, in front of the rectum, only in men. Thus prostate cancer is a disease that is unique to men. The classification of the disease is adenocarcinoma, which means it is associated with a gland. The cause of prostate cancer is not understood, in spite of intensive research.

The prostate gland and the nerves that surround it are a part of the male reproductive system. The gland's functions are to store and secrete part of the seminal fluid, and the prostatic muscles ejaculate the semen upon orgasm. The nerves surround the gland, and are associated with achieving erection. Any treatment that destroys or damages these nerves may render the patient impotent to some degree. Prostate cancer may also cause other difficulties with sexual capability, such as inability to achieve an erection (erectile dysfunction or ED), painful ejaculation or orgasm, and blood in the semen.

The normal size for the prostate is about the size of a large walnut, but size can vary and usually increases with age, sometimes due to "BPH," or benign prostatic hyperplasia. This condition can cause the symptoms of restricted urinary flow, because the enlargement of the gland presses inward and compresses the urethra, the tube that carries urine from the bladder during urination and semen during ejaculation. The urethra goes centrally through the prostate gland itself.

Becoming an Informed Patient

The symptoms of BPH are reduced urine flow, intermittent flow, increased frequency, interruption of sleep to urinate (nocturia), painful urination, and even complete urinary blockage, which is a serious condition that requires immediate medical attention. All these symptoms may also be indicative of prostate cancer, therefore it is important to seek competent medical attention if they appear.

Unfortunately, prostate cancer can also develop with absolutely *NO* symptoms. A man can have the disease and feel completely normal and healthy until the cancer has reached an advanced stage. What happens is that microscopic cancer cells escape the prostate "capsule," usually following a path provided by the nerves, to the seminal vesicles, lymph nodes, and then the bones. This process is called metastasis. This is why the yearly examinations for PSA and the DRE (explained later) are so important; you want to find it early, *NOT* late.

When the disease reaches an advanced stage, other symptoms will start to appear. Some of the most common signs, besides those described above, are what is described as "bone pain," usually in the spine, ribs, or pelvis. These can vary from dull persistent aches, to sharp and debilitating shooting pains. If the spine is attacked, it can cause pain, leg weakness, and urinary and fecal incontinence (bowel leakage). As the disease progresses, any or all of these symptoms can increase. In "end-stage" cancer, when a cure is no longer possible, maintaining some quality of life becomes the primary goal of the doctors and caregivers. Along with bone pain, the risk of fractures becomes a danger, so care must be taken to avoid falls or other stress incidents that could cause them. The primary goals of therapy in end-stage prostate cancer are to slow the progression of the disease and to reduce or control the pain associated with the bone metastases.

A male has approximately a one in six chance of developing prostate cancer during his lifetime. This is a fact, and may be verified by referring to any number of medical treatises. If there is a history of prostate cancer in the family, the chances increase significantly. According to "Us Too," a Prostate Cancer Support Group: *"A man with one close relative with the disease has double the risk, with two close relatives his risk is five-fold; and with three, the risk is 97 percent."* [1] These odds are not those that you would want to gamble your life with! Make no mistake; prostate cancer is a killer. Statistics show that each year almost 250,000 men are diagnosed with the disease. About 30,000 men a year die because they either did not discover the cancer until it had metastasized, or they waited too long to take any action.

It is a fact that as a man gets older, the chance of developing prostate cancer increases. But it can and does strike adult men of almost every age. One factor that leads to the possibility of simply ignoring any warning signs is that prostate cancer usually does not start to evidence itself until after the age of about forty. It is probably true that the most dangerous period—in so far as prostate cancer is concerned—in most men's lives, is between the age of 40 and 60.

In my opinion, this is due to the fact that this is the time of life that is the busiest and most productive in terms of our daily lives, and working at our jobs. Many men are "too busy" to take the time out of their active schedules to have their yearly physicals. Most men are at the peak of their physical and mental capabilities, and maturity. If they are reasonably healthy, the last thing that they are thinking about is the *"remote"* possibility that some disease may be lurking—or perhaps already growing and doing its dirty work on them. You may say:

"I don't feel sick, so why worry?"

But sometimes there are *NO* early symptoms of prostate cancer! It is a *"stealth invader!"*

Prostate cancer is slow to develop in *most* cases. However, this is not always true. Some prostate tumors develop comparatively rapidly, and this is why PSA checks and digital rectal examinations (DRE's) should be a mandatory part of every man's yearly physical exam. For those who don't know, PSA refers to "Prostate Specific Antigen," which is a blood factor that indicates the presence of prostate cancer cells, and is a primary diagnostic tool. For an excellent explanation of PSA see:

<http://malecare.com/prostate-cancer_77.htm>,

Malecare.com, which also explains the use of "free PSA" in diagnosis. You may have to ask for the free PSA test; it is a separate analysis but can help in the diagnosis of prostate cancer. A qualified laboratory can run these analyses using a standard blood sample from the patient.

The DRE is another factor that causes many men to forego the yearly checkup (or this part of it). The procedure requires that the specialist insert a gloved lubricated finger in the rectum to feel (palpate) the prostate gland. Some men reject the idea of this. They should not, because abnormalities such as hard spots or lumps may indicate a tumor. When lumps or nodules are felt, an experienced urologist, preferably one that specializes in prostate cancer, should repeat the DRE. Under no circumstances should this test be relegated to an assistant or nurse! The doctor who may be the one that treats you (through any treatment method) should do it!

This first chapter is directed to those men who are concerned about prostate cancer, or have become aware of it because they, a family member, or friend has been diagnosed. The purpose here is to send a wake-up call and to make sure that the seriousness of this disease is understood. As stated previously, it is the second leading cause of cancer death among men in the United States.

Get the yearly checkup, including the DRE,
and track your PSA!

If the PSA is under 2.0, *AND* there is no hard spot or "nodule" felt by that doctor's probing DRE finger, good for you! You may have dodged a bullet, at least for now!

Start a record on a sheet of paper. Tape it up somewhere that you will see it every day. You *want* to be reminded of this, for the good of your family and yourself! Put the date down that you had the PSA and DRE, and record the PSA. This is the start of a record that you *MUST* keep and maintain, with a yearly reading mandatory! If you are over 50, do not defer this. If there is a rising trend, get your PSA checked every six months!

When you record your PSA the next time, if there is a significant increase, you should let your doctor know that you are concerned, and *ask your doctor for a referral to a urologist.* Do this even if the doctor has *NOT* recommended any action. What is a significant increase? If the reading adds 0.75 ng/mL or more, it is a concern; if it adds one unit from the year before, (like from 2.0 to 3.0), that is certainly significant.

There will always be small variations, because lab results are inexact. So long as the reading is less than 2.0, you are most likely OK, but this is where the DRE becomes important, as will be explained later. Infections may also cause a PSA increase.

Another fact that has been pointed out in other prostate cancer books is that the PSA reading may be increased by activities such as horseback riding, riding a bicycle, having sex, and the probing finger of the doctor doing a DRE! Many doctors are unaware of this. Therefore, in your own interest, abstain from sex, weight lifting, and bicycling type activities (anything that exercises or presses upward toward the prostate) for at least one week before your blood test, and make sure that your blood is drawn *BEFORE* the doctor does the DRE.

Tell him what you want, if he starts to do the DRE before the nurse draws your blood. If he acts like he doesn't like it, so what? You are the paying customer (and now becoming informed patient), right? Insist! PSA's vary a lot anyway, so in this way, your PSA reading will more likely be a correct indication, and not skewed higher because of external (to the prostate) events.

In other words, if the doctor did the finger exam just *BEFORE* drawing blood, or a bike ride was taken before the test, or there were sexual activities recently before the PSA blood was taken, it could possibly cause an elevated PSA. If this is acted upon, the uncomfortable (putting it mildly) prostate biopsy will then have to be endured. In addition, the frightening prospect that cancer might be found will be experienced.

If a high PSA is reported, a good plan is to have the PSA rechecked (using the above precautions and waiting four to six weeks) before proceeding to a biopsy. Most doctors will usually schedule a round of antibiotics before scheduling a biopsy, to rule out an infection that might have caused the increased PSA.

A hard spot or nodule found by the DRE is a sign that something may be wrong. It *COULD* be an indication of a tumor, *EVEN IF YOUR PSA IS LOW!* A suspicious DRE finding by your primary care physician should be followed up by the examination of an experienced urologist.

Why bother to have the DRE if a positive finding is going to be ignored?

This I learned the hard way. In 2005, my PSA had increased from 3 to 4 in one year, and my doctor reported that he felt a small nodule. He did NOT recommend any action! He mentioned it, but it was as if he was mentioning an unimportant circumstance in passing. I was ignorant of the significance of the nodule.

I went away "fat, dumb, and happy," thinking that I was in good shape! Not only was the significant increase of the PSA an important warning sign, the nodule was a double warning! It turned out that he is one of many physicians who believe that as long as a patient's PSA is under 4.0, there is no cause for alarm. This, I know now, is just plain wrong! In July of 2006, the same probing finger found the nodule (larger) again, and when my PSA result came in it was 5.0. **THIS** time he immediately wrote out a prescription referral for the urologist! I had crossed his "magic" 4.0 threshold, you see! Thus I unknowingly "lost" a year in my battle. For a year the dreadful thing had continued to grow, during which time other warning signs such as having to get up one or two times at night to urinate, and episodes of "weak stream" appeared. I was an uneducated, uninformed patient, and I missed an opportunity to win the first round in my fight! This doctor is no longer my GP.

General practice doctors are not infallible, and neither are specialists! (See the story later in this chapter.) Granted, they have the advantage of extensive medical training, but most general practitioners have too many patients, and it is almost understandable that they must use general guidelines in doing routine physical examinations.

We—as the patients on the receiving end—should educate ourselves and understand the tests that are supposed to provide the early warnings that something may be starting to go wrong, or provide evidence of such. These tests may ultimately decide our quality of life for the remainder of our time on this earth, or perhaps even if we will be here at all. Understanding of the tests and of the conclusions reached by the doctor or other medical specialist who ordered them is absolutely necessary! Therefore it is imperative that you become an empowered, educated, and informed patient.

Becoming an Informed Patient

I hope that you are beginning to understand that you and you alone are ultimately responsible for the medical decisions and actions taken that affect your life and the way you may be able to live it.

It is you, who must sign those hospital waivers, not the doctor, is it not? It seems that we tend to blindly accept the decisions and recommendations from physicians. This acceptance has been ingrained into our upbringing, since doctors (in their impressive white coats) have always been authority figures. Yes, and we were always told: *"The Doctor Knows Best!"*

If you *choose* to be uneducated and uniformed about the medical issue being discussed and diagnosed, you will probably accept the doctor's first diagnosis and recommendation. This you should **not** do! You must realize that this is only his **opinion,** in which he may have a vested interest!

If, on the other hand, you study the issue and become more informed, you can better understand the disease and whatever tests or treatments are recommended. You will know to get at least one "second opinion." Two or more "second opinions" are even better! You will also learn the potential outcomes of proposed treatments, along with, hopefully, some of the advantages and possible drawbacks. And you will know what questions to ask in order to make intelligent decisions. You may, in fact, during your study and searches, find an alternative treatment that was not recommended by a particular doctor, but one that better fits you as a person and your life as you want to live it. *I did. I found "The Beam!"*

Others have also done this. Here is a true story that will illustrate what I am trying to get across in this chapter far better than anything that I can write. This is from an Internet friend that I met on the Yahoo Prostate Cancer Support Group, Mr. John Shuey. He posted this on the forum July 22, 2007.

9

It says it all, and is entitled simply:

"The Doctor Said ...."

"Just under fifteen years ago my wife and I sat in a small room at Georgetown Univ. Hospital and listened to the chair of Pediatric Cardiology describe the several things wrong with our newborn son's heart, and then add, "Given what we know today, the best thing for you and your son would be to take him home and allow him to die."

"What about a heart transplant?" I asked.

"They don't work in babies," she said. "Even if you were lucky enough to get a heart, and if your son did live, he'd be a 'bubble baby', unable to go to school or play with friends."

Why am I telling this story to this group? Let me finish and you'll understand.

After several days of research, including phone calls to dozens of specialists around the country, we opted to take our son to a hospital on the west coast that specialized in infant heart transplants.

Today, he is 15, soon to be a high school freshman, an honors student, and active in numerous clubs, etc.

The point is that even some of the brightest and best can be uninformed and/or prejudiced. We all need to take the time to figure out what's best for us and not worry what anyone, even doctors, think.

Oh...and there's a second point too: The hospital on the west coast was Loma Linda. And that's where....in spite of some reticence on the part of my urologist.... I've decided to go for Proton Therapy." [2]

Note: Loma Linda University Medical Center was where the very first infant to infant heart transplant was done, in 1985. LLUMC remains on the cutting edge of advancing medical practice. Mr. Shuey did become a proton prostate cancer patient there in 2007.

Becoming an Informed Patient

Thank you John, for this enlightening evidence, which proves that we must all strive to be better informed when we or our loved ones are diagnosed with diseases or illnesses, and that the doctors may *not* always know best.

This informed approach is of vital importance in any battle with the Big C! When one leaves the office of the general practice doctor and enters the more rarified atmosphere of the specialist, then understanding the significance of tests and results is even more important. If you have questions, the specialist is supposed to be able to provide the answers, and help you understand the significance of them and what they mean to you.

Is that not why you are going to the specialist?

Therefore, prepare for every session with a written list of questions, and please take someone along with you so that later you can compare notes and recollections of answers and statements. You might even consider a small tape recorder to refer to later.

If the urologist (or other specialist) does not want to answer your questions—***and there are NO questions too dumb***—or if he is not forthcoming and seems not to have the time or inclination to help you understand, remember that you are the patient (and paying customer). If you are doing your homework, and are truly doing your best to become an empowered, informed patient, and have any questions that are not answered, or if there are alternative therapies that the specialist won't discuss because his recommended treatment is the *"only"* treatment, please remember that you have the right, ***NO, the obligation to yourself,*** to leave that office and seek competent assistance elsewhere.

There is no reason to feel guilty about this, nor does it have to be confrontational with the doctor. It can be done by telephone, writing a brief letter (best), or simply saying to the receptionist on the way out:

"I have decided to find another doctor, please hold my records until you are asked to forward them."

There are such doctors as described above. But please believe me when I tell you that there are also many very skilled doctors in almost all larger communities who will welcome an educated and informed patient, and are perfectly willing to work with such a patient every step of the way! And this is the way it should be!

The doctor–patient relationship should be one of trust and mutual respect; and what better way is there to develop that relationship than being able to understand and frankly discuss the issues and alternatives in various diagnoses and proposed treatments.

Dr. Charles M. Balch, Professor of Surgery and Oncology at Johns Hopkins and a recognized author of more than 500 published articles, has been quoted as follows: *"I truly believe that an informed and educated cancer patient will find a way to get the best care ... and live better and longer."*

We must each decide which treatment is best for us as an individual. Careful study will show you that *ALL* the "modalities" will yield at least similar chances of "cures." When a treatment decision is finally made, you will rest easier knowing that as an empowered, educated, and informed patient you have done everything that you can to understand the disease, the treatment, and—most important of all—the potential outcomes of that treatment. Just remember, you will have to live with that decision the rest of your life, so you must do what you truly believe is the right thing for you!

It is not easy, starting on this journey, especially in the initial stages of fear, frustration, and differing opinions found following the dreaded diagnosis. It takes courage and fortitude, and perhaps the stubborn idea that you can beat this thing called prostate cancer.

Becoming an Informed Patient

First, resolve in your mind that you will take the time and expend the effort, whatever it takes, to learn all that you can, from any source, about the disease and every treatment possibility. The more you learn, the easier it will be to find a treatment that you want for yourself.

If you are newly diagnosed and feeling confused and overwhelmed, just recognize that this is normal. Take your time, and concentrate on one thing or treatment at a time. Make notes and start an organized notebook to refer to. If you take it slow and easy, things will gradually become less confusing. My personal "learning curve" to become less confused and minimally knowledgeable was about three months. Your time might be more or less, depending on your time and/or resources.

Libraries are great, and there is usually a research librarian that will assist you. The Internet, in spite of the warnings "be careful about what you find," (and you should!) is a wonderful resource, and search engines like Google and Yahoo are valuable tools. Use them!

Second, and this is very important, recognize that **"You Are Not Alone,"** and take advantage of the experience of the men who have already traversed this fearful pathway, and survived. The above quotation is a reference to Terry Herbert's Internet website, YANA (*<www.yananow.net>*), that should be one of your first places to start. There is a wealth of information that will serve the newly diagnosed prostate cancer patient well, plus much useful data for those further along on this journey. In addition, there is a section called "Mentor Experiences" that detail the actual experiences of prostate cancer patients with the various treatment modalities.

Terry Herbert is "one of us." After diagnosis, he chose "Active Surveillance," and has been fighting the good fight ever since. Terry's expertise and advice to any who seek it is always available. He modestly says: *"... I was diagnosed in '96: and have learned a bit since then."*

Here is another very pertinent quote from Terry Herbert: *"How many cases of 'cancer' are in fact misdiagnosed at present because the additional tests that are currently available to verify the diagnosis are not used. To quote* [well known cancer specialist] *Dr Charles "Snuffy" Myers: 'As a physician, I am painfully aware that most of the decisions we make with regard to prostate cancer are made with inadequate data.'*

It doesn't have to be that way, but to change things requires all of us to be more demanding in our approach, not to just accept what a doctor tells us, especially as he may have a vested interest in what he is advising. We simply need to be more pro-active—all of us."

There are local support groups like "Us Too" and others that you should locate and join. There is no substitute for the personal experiences of those who are veterans of the battle.

This thought leads directly to another great site to visit, "Phoenix5", (*<www.phoenix5.org>*), started by Robert Vaughn Young (1938-2003), who was another veteran of the battle. He provided the following on one of his pages:

"To My Fellow Men ..."

"In every struggle, the only ones who can truly grasp your fear, your pain, your grief, and your stamina that may sometimes fail, are those who share the battlefield with you. It is no different when the enemy is prostate cancer and the fight is for your integrity as a man as well as your life.

...We are hurt and confused but there are some others who know. Don't be ashamed to hurt. The wound is deep. But it need not be fatal. Nor are you alone. There are many of us who know the feeling and some of us who want to help the rest. That is the basic message of Phoenix5."... [3]

~ Robert V. Young

"You Have Prostate Cancer!"

S O MY G. P. SAID: "I'm going to write you a referral to a urologist; you need to have your prostate checked out." This was after an increased PSA level from 4.0 to 5.0 in one year; and the DRE found the nodule to be still there, slightly larger and harder. When I asked about the significance of the nodule, he told me that sometimes a nodule could indicate the presence of a tumor.

I stood there struck dumb with this statement. I did not even have the presence of mind to ask him why he did not tell me that last year! I knew that tumor meant cancer—my grandmother had died from a tumor the size of a grapefruit in her stomach. The doctor then told me that it might be nothing, but that the rising PSA and the nodule were "suspicious." This was the beginning of a constant series of days full of the frightening thought that I might have prostate cancer, the "Big C!"

I had almost two weeks until my appointment with the urologist. Finally the day arrived, and my visit to this specialist was brief and to the point. He simply said to me that I needed to have a biopsy of my prostate, which he would do there at his office. He told me that he would do an ultrasound inspection, and use hollow core needles to take the biopsy samples. He said that the simple procedure would be *"... a little uncomfortable, about like drawing blood."*

Looking back on this statement made me wonder why this doctor lied! I finally came to the conclusion that he himself must never have been subjected to the procedure, so how could he possibly have known how painful it would feel to me?

He gave me an antibiotic to take before the procedure, and explained that the needles would be injected into the prostate gland through the rectal wall, thus there was a slight chance of infection, hence the antibiotic. He would use ultrasound to guide him while taking the samples.

When I went ahead with the procedure, it was more than "uncomfortable," and every time (fourteen) the needle gun snapped, it hurt like hell! The doctor's report, that I later read, said: *"Patient tolerated the procedure well."* Well, I did not! Each *SNAP* felt like a huge rubber band being stretched back to the limit then released against something well inside my body. I thought with each *SNAP*, surely that will be the last! I counted each and every one!

My recommendation is that if you have to have a prostate biopsy, tell the urologist that you want some strong pain medication or a local anesthetic. Make sure that he takes at least twelve, preferably twenty biopsy core samples. Tell your doctor you prefer more samples rather than few. Most doctors now stop at twelve; twenty core samples provide an improved chance of finding elusive cancer cells. The problem is that the beginning tumor (if it is early in the formative stage) or even if it is later in the process, may be missed by the biopsy needles. Cancerous cells may be few and scattered, thus easily missed. It is a bit like the proverbial "needle in a haystack," and "false" negatives (finding no cancer with the biopsy when in fact the cancer does exist) are not infrequent. If there is a high PSA reading, there is the presumption of prostate cancer after infections are ruled out, and repeat biopsies may be required.

There are additional recommendations spelled out in other books and reference material that you should investigate during your efforts to become an informed patient. Bob Marckini's book has an excellent section on this, and more information can be found on the Internet.

The biopsy served to escalate my anxiety. Now I was waiting to find out if the samples would show anything cancerous. I still knew very little about the disease, and tried unsuccessfully to put it out of my mind by working on a genealogy project that was ongoing. Somehow the "What if?" question seemed to always creep in, and I could not seem to concentrate on my work. I kept telling myself that I felt fine, how could anything be wrong? In about a week the phone call finally came from the doctor's office: I was to meet him in his office the next day. I did not sleep much that night.

When we met the following day, I had the feeling of dread that had been steadily growing worse since the biopsy. This doctor was very curt, and did not mince words: *"You have prostate cancer. Your Gleason's Score is 8, and that indicates an aggressive cancer."* These words, whether bluntly stated or delivered in some other fashion, are shocking and numbing! After having gone through the procedure of a prostate biopsy, the patient will have recognized the possibility that the cancer may be there, especially if he is trying to become an informed patient. Still, there is that old mentality that:

"It can't happen to me!"

The urologist told me that he recommended "Seed Net" cryosurgery that he would perform, but that my prostate was too large to do the procedure and would have to be "shrunk." I later found out that it was measured during the biopsy at 56.6 cc, which is about double the normal size. The shrinking would be done by a Lupron injection, which was hormone therapy, and would *"shrink the prostate, and keep the cancer cells from growing in the mean time,"* he said.

He gave me a videotape describing the cryosurgery, and told me to come back in three days for the Lupron shot, and in a week to let him know what I had decided. I remember sitting there trying to think of some question to ask, and being in such a state of shock that I could not think of a single one. As I got up to leave, he had to remind me to take the tape and to make the follow-up appointments. Somehow I did this and drove home, but do not remember even doing it. What was I going to tell my wife to lessen the blow? How could I somehow make her less afraid? These and other questions came and went rapidly. It turned out that I had to tell her nothing; she took one look at me and knew! All we could do was to hold each other tightly. I blurted out what the doctor had said about the "Gleason's Score," and then realized that I had no clue what that even meant. That was the first time in my life that I had even heard of it!

I think that it was at that moment that I started to realize that I needed to know more about this evil thing called prostate cancer. Once again I did not sleep much that night.

The next few days were not good. Shock and fear had taken their grip on me. All the dire possibilities seemed to rotate continuously through my mind. I could not seem to concentrate on any one thing long enough to seriously consider it. My wife and I talked, but looking back, I cannot remember the conversations at all. I do remember that she did her best to comfort me, and reassured me that we would get through this together. I went for the Lupron shot without doing any research on this hormone therapy at all! *I was numb with shock, and was the perfect example of an "uneducated, uninformed patient." I was in no condition to make any decisions that might affect the quality of life, as I might want to live it!* The lesson here is to never make hurried, un-thoughtful medical decisions if the situation is not immediately life threatening.

It took almost a week before some semblance of normality seemed to prevail. I do remember getting my will and trust documents out of the safe and trying to determine if anything needed updating.

The Gleason Score question seemed to be uppermost in my mind, so I started investigating it. Named after the doctor who developed it, the Gleason Score is a means of classifying the grade and aggressiveness of the cancer, based on microscopic investigation of the core samples taken during the biopsy.

There are two dominant patterns identified, a well differentiated one, and a poorly differentiated one. The well-differentiated pattern is the one consisting of normal, healthy prostate cells, and the poorly differentiated one is the pattern consisting of the uncontrolled, mutating cancerous cells. The pathologist observing these cells under the microscope determines the most prevalent pattern and assigns it a "grade" based his determination of the amount the dominant one.

The following diagram is found on "Phoenix5," a very informative and useful Web site, but the diagram is now prevalent on many sites, including "The Prostate Lab," "AzTecFreenet.Org," "Malecare.com," and others. As such, and due to the fact that the diagram is Dr. Gleason's work with no apparent copyright, it is reproduced here for educational purposes with no further reference.

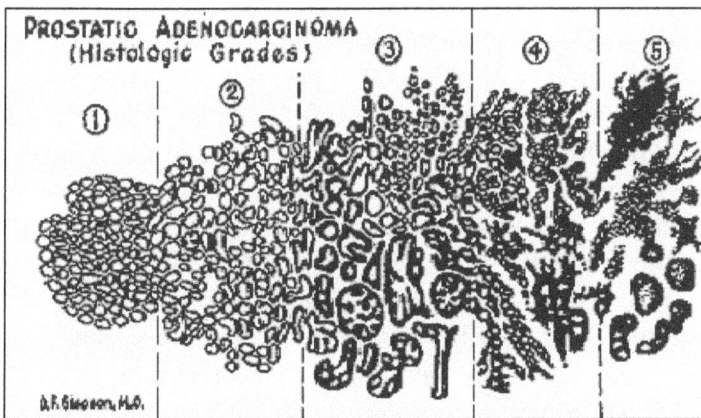

"The Gleason score indicates the aggressiveness of the prostate tumor. Patterns 1 and 2 are well differentiated; Pattern 3 is moderately differentiated; and Patterns 4 and 5 are poorly differentiated. ... The Gleason score is written as the sum of the two most prominent ... patterns. So a Gleason score of 2+3=5 has a dominant well-differentiated pattern (i.e., pattern 2) and a less dominant moderately-differentiated pattern (i.e., pattern 3). A score of 4+3=7 means that a poorly differentiated component (pattern 4) is dominant. If 95% or more of the tumor is composed of one pattern, the corresponding number is counted twice; thus, a wholly moderately-differentiated tumor would be scored 3+3=6." [4]

This shows the complexity of the process, and the need for expert analysis. As you can see, this "grading" is extremely subjective and dependent on the expertise of the pathologist who is performing the analysis. A good site to start reading about this is "The Prostate Lab," the site from which the above picture and quotation was taken, and "Phoenix5," where that site was first found.

Then, I suddenly remembered that I had another appointment with my urologist! This seemed to force me to settle down and begin to think. The doctor had said to view the video, and I had not even done that!

What was *"SeedNet?"* What was *"Cryotherapy?"* My wife and I looked at the tape together, and saw that the procedure (which looked very simple in the animated tape) consisted of inserting hollow tubes in the prostate gland and circulating super-cold (cryogenic) gas through the tubes into the prostate, until an "ice ball" was formed in and around the gland. Then this procedure was repeated. This freezing was supposed to kill the cancerous cells. There was no mention of side effects or potential problems in the videotape at all. It was definitely a "selling tool," and a "hard-sell" at that!

At this point I was in the state of mind that many, I am sure, are in following the diagnosis of prostate cancer. I just wanted to accept the doctor's recommendation, and not worry about making a decision.

After all, he was the doctor, and the "doctor knows best," right? And my GP of thirty-five years must have confidence in this man, since he sent me to him, right?

I truly believe that if I had not already been given the four-month Lupron hormone shot, and the cryotherapy could have been done sooner, I would have gone ahead with it just to *"get it over with!"*

How many radical prostatectomy operations must have been performed on men in just this state of mind! I have a close personal friend who was diagnosed just a couple of months before I was, and he opted for the surgery immediately after the doctor recommended it, "Just to get rid of the thing!" He did no research, and never knew of the proton beam or any other option at all! His urologist told him that surgery was the "best" option for him, and so my friend took him at his word! Time will tell if it *was* the best.

Another point: what is it with some doctors, especially urologists, that seem to want to use the "shock effect" on patients when imparting the news that the patient has cancer? I know in all the education that they received, they *MUST* have had some training on how to impart bad news. If not, then common sense and a bit of simple humanity should suffice to know *NOT* to call a patient on the telephone at 7:00 p.m. in the evening to tell them they had cancer.

But I received an email from a distraught young lady who related that this is exactly when and what a urologist did to her 77 year-old father. You know he and his family got no sleep that night!

Why not call in the morning and ask the patient to come in to the office? Or if the bad news must be relayed over the telephone, call in the morning, and lead the patient to the bad part by first attempting to put him at ease by imparting some good fact about the situation.

Regardless of how the diagnosis is provided, I am firmly convinced that anyone facing *any* form of surgery (or any other treatment) should make an attempt to learn as much as possible about the procedure recommended and potential side effects. *THEN* the patient should investigate other alternative treatments or procedures that are available for the disease or condition. Also, at least one second opinion should be obtained from a recognized specialist in the field for each different treatment option. Two or more are even better.

The same is true for the different "modalities" or treatment methods for any cancer. When you seek out or receive a recommendation for a particular type of treatment, you will find that these are what they are called, *opinions.* From these opinions *you must decide for yourself* whether or not the recommendation is the best way for you.

In his book, Bob Marckini relates how, in seeking a treatment for his cancer, various doctors told him that he was a "poster boy" for radical prostatectomy, brachytherapy, cryotherapy, and conventional radiation. These were *opinions* from different specialists. In my own journey I had the very same sort of experience.

Therefore, seek out the *very best* specialists that you can find in order to receive the best opinions that you can possibly get! You may find that you have to travel to get a true expert in the field. If that is what it takes, that is what you should do, if you possibly can. You truly need the experience, expertise and information, and the decisions reached as a result of these consultations will probably affect your quality of life for the rest of your days.

Chapter Three

Research and Doctors

A S AN ENGINEER, I had become familiar with cryogenic (super-cold) liquids starting in the 1960s. Liquid nitrogen and that coldest of elements, liquid helium (minus 452 degrees F), were routinely used as coolants. Liquid oxygen and liquid hydrogen were the rocket engine propellants used in both the Centaur and later the Shuttle vehicles that I spent most of my NASA career preparing for launch. So this method of treatment had some appeal at first, since I was familiar with super-cold fluids. I decided to "Google" SeedNet Cryotherapy. This first step was the beginning of my education about the treatment of prostate cancer.

When I started my research, I came to realize the true benefit of the Internet and search engines like Google. Of course reading books and papers in libraries can do the job, but hundreds of hours can be saved by the use of the computer and the Internet! I entered "SeedNet Cryotherapy" as the Google search object, and hit "enter." There were 495 entries found in 0.04 seconds! The following is a typical result:

"... Oncura's SeedNet system is a minimally invasive treatment for prostate and renal cancer. Easy to learn and simple to use, SeedNet provides optimum accuracy, efficiency and control in the cryoablation process. The SeedNet system employs ultra-thin 17-gauge CryoNeedles as opposed to outdated 3 mm cryoprobes. These slender needles are specifically designed for direct percutaneous insertion, eliminating the use of an insertion kit, and can be added, removed or repositioned within seconds causing minimal trauma. CryoNeedles needles generate precision iceballs that combine to create a frozen region

with uniform lethal temperatures. The SeedNet system is specifically designed for use by urologists and presents an innovative alternative to prostatectomy and radiation treatments. Based on the brachytherapy setup familiar to many urologists, this procedure is easy to learn and simple to use. Ultra-thin CryoNeedles guided through a template are percutaneously inserted into the prostate and generate cancer-killing iceballs that destroy prostatic tissue." [5]

This answer was specific and on point, but then a new set of questions arose: What is brachytherapy? What is involved in a prostatectomy? What are the various radiation treatments? What are the advantages and disadvantages of them? Are there other alternatives? The urologist who did my biopsy had explained none of these!

I was suddenly plunged into a sea of unknowns. I knew that my task was just beginning. But first, I had to prepare for my upcoming visit to the urologist, and make my list of questions to ask! Here is what I came up with:

1. Can I get copies of my biopsy sent to another lab for a second opinion?

2. Did he feel my tumor (was it palpable) on both sides?

3. How often will you check my PSA, and how will you track the progress of the Hormone Therapy?

4. What was the size of my prostate in cc's?

5. What are the other (besides hot flashes) side effects from the Lupron, both short and long term?

6. Why do you recommend cryosurgery for me? What are the pros and cons of cryosurgery vs. other treatment alternatives?

7. Will I have to go off of my Plavix for any of these treatments? (I take Plavix because I have had a stroke.)

8. Where are the recognized "Centers of Excellence" for the cryotherapy treatments?

In the time remaining before the appointment, I found the answers to some of these myself, but asked the doctor anyway. On the day of the appointment my wife went with me, both for moral support and to make notes of the answers to my questions. The answers in general to the above questions are given below as best as we can remember from memory and recorded notes.

During this session, it must have become apparent to him that I was not prepared to immediately agree that his "SeedNet Cryotherapy" was my choice, because the doctor asked me if I had made my decision yet. After I said that I had not, and was still investigating alternatives, he became almost non-committal and seemed irritated that I was asking questions. I tried to explain that I felt that I was a reasonably intelligent person who, along with research, could understand whatever medical explanations he gave, but it seemed to do no good. His answers:

1. He could not arrange to send the biopsy slides to another location; I would have to arrange that. He offered no help on how to do it.

2. He said that he felt a nodule on both sides. (My GP had said the nodule was only on the right, and the biopsy results were mainly on the right.)

3. The PSA would be rechecked in two months.

4. The prostate was 56.6 cc's, about twice normal size.

5. He stated that there were no long-term effects from the Lupron shot; hot flashes were the most common. (Absolutely no mention of the loss of sex capability, muscle loss, or the mood swings that I later experienced! It was about at this point that he seemed to become irritated at my questions, and offered no additional information.)

6. He said that cryosurgery was best because of my "old age," and that regular surgery would not "get it all" because of the Gleason 8 score from the biopsy.

(These statements made a little sense after I became better informed.) He then said that the cryo *was* repeatable *"if the cancer comes back;"* and that radiation from seeds *"may cause bad effects on the bladder."*

7.   He told me that I would have to go off of my Plavix and vitamin E ten days before the surgery.

8.   The only other locations that he would recommend (for cryosurgery) were Mayo Clinic in Jacksonville, Florida and perhaps a Dr. Moffett in Tampa, Florida. (I did not pursue either of these.)

He then suggested that I come back in about two weeks if I had additional questions or if I had any problems. Then I asked one last question: "If I do nothing, what is the prognosis?" His answer: *"You will probably be dead in about two years."*

I became very focused after that.

Much later in my research, I found that once again this doctor either was not being truthful or was just ignorant. If untruthful, he was resorting to a pressure tactic that was most unconscionable, in an effort to have me make a quick and unthinking decision. The truth was that I probably had from five to ten years, perhaps more. But back then I did not know, and he was the doctor!

Following that visit, I investigated whether or not the cryosurgery that he had recommended was possibly the correct course of action. I read the information of Dr. Gary Onik, a recognized authority and "artist" in cryotherapy of the prostate, whose published work led me to the idea that "focal" cryosurgery may indeed be the right treatment, hopefully to spare the nerves on the left lobe (I wanted to avoid becoming impotent). I faxed my records to his office in Celebration, Florida (south of Orlando and only an hour away).

Dr. Onik reviewed my records and sent me the information that he could treat me, but that another more extensive "mapping" biopsy would be required first. I was told that the cost of the initial consultation, the biopsy, and the cryotherapy would be $9,500.00 cash, because Dr. Onik had "opted out" of Medicare and accepted no insurance. Medicare is my primary insurance, with Blue Cross-Blue Shield as secondary. Dr. Onik referred me to Dr. R. Brunelle in Tampa, who did accept insurance, and he was good enough to correspond with me, and after review of my records, indicated that I was indeed a candidate for cryosurgery if I should decide on that course of action. *This of course was also his specialty.*

Well, I was still in somewhat of the *"Git 'er Done"* frame of mind, and since Dr. Onik's office and the Florida Hospital was an hour away, and I hated driving to Tampa on I-4, I seriously considered going ahead with Dr. Onik. He was a recognized "artist," and people came from all over the world to have him do their treatment, paying cash. This impressed me! But again, the fact remained that I had the Lupron, and Dr. Onik indicated that I should wait at least three months after administration of the shot to have the procedure, so I continued my research.

Then I read of the surgical technique called "Da Vinci Robotic Surgery," that was less invasive and traumatic to the body than radical prostatectomy, with less blood loss and a faster recovery time. I found a doctor in Winter Park, Florida who had done many of these, and had a good reputation. I made an appointment, and my wife and I went to see him, taking along all my prostate lab reports and records. This urologist was excellent and thorough, taking the time to explain to us that I was not a candidate for any kind of surgery, because of my age and the fact that I had suffered a stroke. **Looking back, I now thank God that I am old and had suffered a stroke!** Otherwise, I might have opted for this surgery!

Even if I had been a candidate, I later found that there were too many potential problems for my own peace of mind. Others may find robotic surgery acceptable. Several good articles on this are available on-line.[6]

This urologist, a surgeon, recommended radiation by intensity modulated radiation therapy, (IMRT), which is an advanced mode of radiotherapy that utilizes computer-controlled X-rays to deliver precise radiation doses to the cancer site, at different angles to **reduce** radiation damage to nearby organs. Here, "reduce" is the operative word. As explained elsewhere, X-rays or photons do not stop, and in continuing through the patient's body, deliver harmful radiation to other body tissue and organs.

Since I had not thoroughly investigated radiation therapy, but having found out some of the side effects of it, I was not ready to pursue this yet and did not make an appointment with the recommended radiation oncologist.

I believe that my indecision in this process was the best thing for me, because I proceeded to read and study everything that I could find about prostate cancer treatment methods (modalities, as the doctors called them). Even though I had "jumped" into the decision to go ahead with the hormone treatment with Lupron, that decision allowed me the time that I needed to do my research and study. Then I found the book *"A Primer on Prostate Cancer"* by Strum and Pogliano. I recommend this book to anyone, patient or family member, whose life has been affected by prostate cancer. However, it is a reference book—written by a doctor—and parts may be difficult to understand, but it is worth the effort.

However, before you buy or read ANY other book, I highly recommend Bob Marckini's wonderful treatise, *"You Can Beat Prostate Cancer And You Don't Need Surgery To Do It."* I did not have his book at this point in my struggles, but if I had, it would have made my research much easier and more manageable!

I did additional research, but after about a month, still had reached no definitive decision on the course of treatment. I read all that I could about cryotherapy, then brachytherapy, which is radioactive seed implantation, both with and without additional conventional radiation, and radiation (X-ray) treatment alone. I did no further research on radical prostatectomy. I did some additional investigation on brachytherapy, and actually corresponded with the Radiotherapy Clinic of Georgia, in Atlanta. This center uses a combination of seeds and external radiation.

While relatively rare, I found that potential side effects of brachytherapy could be incontinence, urinary pain, rectal bleeding, erectile dysfunction (ED), bladder injury, or rectal fistula. Then I read that the "seeds" could somehow leave the prostate capsule and migrate to other parts of the body, including the lungs and the heart! At first I read that seed migration was relatively rare. Nevertheless the possibility was there, and later research showed that indeed, it was not so rare, and that up to fifty-five percent [7] of seed implanted prostate cancer patients might experience such migration, (depending on the medical study) and that while complications from this were relatively rare, instances of pulmonary embolism had occurred. Some facilities are now using "stranded" seeds to prevent such migration, so there must be some concern.[8] Later, I found that there is a procedure that uses "removable" seeds, (they are inserted for a prescribed period, then removed) but the radiation is more intense, with potentially more severe side effects. I knew that brachytherapy had been used with good success for many, but after all of this, I was almost as confused and frightened as in the beginning. Since I was becoming more informed, I recognized a need to know much more before I finally decided on my treatment. I gave myself a self-imposed deadline of December 30, 2006 to decide. That was about the end of the four-month Lupron period.

Almost five weeks passed since my Lupron shot, and an interim PSA check was done on October 3, 2006. The result was 2.41 ng/mL (it was 5.0 when diagnosed). The predicted hot flashes had arrived with a vengeance. They did not seem to bother me so much during the day, but did disturb me when they awakened me at night.

There was also the complete loss of sexual capability, and another side effect of the Lupron (not told by the doctor) was loss of muscle mass and strength. This was very apparent since I go regularly to a gym with weight machines, and I found that I had to reduce weights to do the same number of reps that I had been doing previously. Another unexpected happening was weight gain, in spite of exercise and no change in eating habits. Later I experienced another side effect: unexpected mood swings and emotional reactions to events. There were other minor effects, none good. Some persisted for several months.

Later, my research found the following by Dr. Steven Strum: *"Testosterone Deprivation Therapy and Its Far-Reaching Implications: If there is any area of PC management that necessitates a comprehensive understanding by the physician, it is in knowing the spectrum of effects of ADT on male physiology. A lack of such understanding deprives the patient of available supportive care that can mean the difference between success and failure in the patient's life. This not only relates to preventing or minimizing side effects due to treatment, but also to the patient's compliance with therapy-- whether he will remain on the medications used in ADT or stop them due to adverse effects. In the early 1980s, I began treating PC patients using anti-androgen therapy in combination with an LHRH agonist as one of the first American collaborators working with Fernand Labrie.*

My observation of patients taught me a great deal about the effects of an accelerated and intensified male menopause. The lowering of serum testosterone to castrate levels, defined as less than 20 ng/dL (less than 0.68 nM/L), resulted in a spectrum of possible signs and symptoms that varied from man to man. Some of these symptoms occurred acutely, while others developed over time. However, all were potentially troublesome, if not aggravating, for the patient. If not treated in a preventive manner, such signs and symptoms can have a negative impact upon the patient's overall health. Except for hot flashes and impotency, **many symptoms resulting from androgen deprivation have been discounted by physicians as being due to old age or due to medical problems such as arthritis or heart disease.** *However, this constellation of clinical and laboratory abnormalities quickly develops in younger men or older men in otherwise good health after the initiation of ADT.* **This clearly suggests that these symptoms are not due to "old age" but are characteristics of the androgen deprivation syndrome (ADS)....** " [Bold emphasis added.]

The thing to be learned from this is that when diagnosed with prostate cancer, you had better find a doctor who *SPECIALIZES* in prostate cancer and who is current with all the latest treatments and drugs. Do your homework. Call the local hospitals and ask for the names of doctors that specialize in prostate cancer. Use the Internet and search for well known prostate cancer specialists. Learn before your appointment, so that you will know what questions to ask!

On October 10, I kept my appointment with my first urologist to take him the results of the PSA as he had requested, and resolved in my mind that if he was not more receptive and forthcoming during this visit, that I would change urologists. I had arrived at the conclusion that whatever choice of treatment I made, it would not be local, but at a major hospital or "Center of Excellence," with a doctor or team that had plenty of experience in whatever modality that I decided to use.

When I went into the office the urologist was as terse as before, if not more so, and only stated that the Lupron was working, since the PSA was about half what it was before. After a brief discussion of some of the side effects I had been experiencing, he asked if I had any new questions. Perhaps he had sensed that I wanted to become more involved in the decisions that would ultimately chart the course of my journey, for the rest of my life. It was very obvious that he considered himself to be the doctor and as such my opinions and questions were not critical or important in so far as the choice of treatment for my cancer. He had recommended the cryosurgery, which he would do, and that was it. My final question to him was: *"If I go to a center of excellence hospital, either in state or out, would you continue to be my urologist?"* The answer was quick and definitive: ***"No! I have my own patients to take care of!"***

I left that office knowing that I would not be going back! I knew that I was becoming more of an informed and empowered patient, and felt a great sense of relief, knowing that I would now have an opportunity to find a doctor that I could work with. I promptly made another appointment with my primary care physician, and when I kept the appointment and explained the situation, he gave me a referral to another local urologist in mid-October.

This experience led me to some serious thought about doctors and treatment in general. This was touched on previously, and will be again, but you and your doctor need to have "full disclosure." In all medical decisions, it's your *right* as a patient to have full information, both before and following any medical treatment or procedure. You need to understand and agree to the decision fully *before* anything is done, including potential side effects, and receive a complete explanation from the doctor, in detail, of results and prognosis following the treatment. You should keep careful records (and copies) of the reports that the doctor prepares each step of the way.

I continued to read and study the prostate demon, and felt that I was becoming less confused, but I was still in need of more information to make a logical decision that would be "best" for me.

During the following days, while waiting for the next appointment, I continued my research. One effort entailed simply entering "prostate cancer treatments" in the Google search engine. The amount of information found was staggering! I sat down at the computer for hours, determined to read as much as I could about the various methods, but eliminating cutting surgery for removal of the prostate.

The more I read the more that I became aware of the need to carefully consider the possible side effects and results of the different treatments. I decided that my number one consideration would be *"quality of life"* because I just did not want to be incontinent, nor did I want bowel problems as a result of damage to the colon, or to be impotent if it could be avoided. One site in particular brought this home to me: *<www.prostate-cancer.com>* listed in tabular form all the potential side effects of the treatments that I had been seriously considering, and they all simply scared me out of my mind! The impotence and incontinence things were in all of them! I became discouraged, angry, very frustrated, and once again confused.

While this site is advantageous in describing the various prostate cancer treatments and side effects, I know now that it *SHOULD* but *DOES NOT* include "Proton Beam Therapy," either as a sub-topic in the "Radiation Therapy" category or as a separate category altogether. *The reasons that protons were omitted are not clear, but during my research I found that most doctors or institutions that did not have proton capability did not mention protons. I found this to be true time after time.*

In this particular late night session, I methodically continued down the list of sites, taking notes as I proceeded. One of the last Web sites on the long list was: *<www.prostatecancerfoundation.org>* that I started going through about midnight on October 11, 2006.

In reading through the information on radiation therapy, I found the following: *"In some centers, **proton-based therapy** is used during IMRT rather than the more traditional photon-based therapy. Although early studies have shown that oncologists may be able to manipulate these beams even more precisely, this technology is not yet widely available. Because the treatment planning with these types of* [PROTON] *therapy are far more precise, higher—and more effective—doses of radiation can be used with less chance of damaging surrounding tissue."* [9]

I remember thinking: "What is this???" It was the first time that I *remembered* seeing, or had been told by any doctor, anything about *"proton-based"* radiation treatment for prostate cancer! ***"Less chance of damaging surrounding tissue..."***

Suddenly I was wide-awake! Then I remembered that I had seen something about this proton thing in the book *"A Primer on Prostate Cancer"* that I had not completely read (it is a rather difficult book to read). I went back and read it, then I immediately Googled "proton prostate" and there was a multitude of sites and information!!

It was almost 1 a.m., on October 12, 2006. I continued to read, becoming more excited as I read! The very first site: <*www.prostateproton.com*> was entitled *"Prostate Cancer & Proton Treatment — What your Doctor may not know, and might not be telling you."* ***WOW!*** Then I read that Loma Linda University Medical Center (LLUMC), in California, had been using proton beam therapy to treat cancer since the early 1990s!

I found the Loma Linda site and spent about half an hour investigating it. And then I read about the other centers in the U.S. using this proton treatment, including Massachusetts, Indiana, and now Texas and Florida. Loma Linda had treated over 11,000 cancer patients with the proton therapy at the time I discovered it! This number is now (April 2008) more than 12,000.

I urge any newly diagnosed prostate cancer patients to check this site and the many others listed to decide for themselves if proton therapy should be considered in the decision process. I thought: *"No cutting, no freezing, and no stabbing! No pain! No radioactive pellets in my prostate! No catheter!*

Minimal side effects! No Recovery Time! Medicare covers it! Comparable history for outcome to other treatments!" I guarantee that this is not the first time that this thought will have occurred to the reader, but it was then that it occurred to me:

"This sounds too good to be true!"

The only drawback seemed to be that if I chose Loma Linda University Medical Center, I would have to travel 2,600 miles to have my treatment, and spend two months away from home! I decided to dig into this in depth. For two weeks I spent most of my time making phone calls to various centers that offer the proton treatment, and continuing to read everything that I could find about it, including literally hundreds of testimonials from men who had received the treatment, most at LLUMC, on the Proton BOB Web site.

In the meantime, (October 16, 2006) I met with my *NEW* urologist for an initial consultation and exam. He was most forthcoming and supportive. I posed the question about out-of-state treatment up front, (Would he remain my urologist?) and he had no problem with it.

His attitude was like night and day from the first urologist! But he did ask: *"Why would you do that, when you could be treated locally with excellent results?"* His DRE found the prostate fairly smooth with one harder area on the right side (the nodule had apparently shrunk). He confirmed that the reduction in the PSA from 5.0 to 2.41 showed that the Lupron was working.

He stated that in his opinion I did not need Casodex or Avodart in conjunction with the Lupron Therapy. He carefully went over the available treatment options with my wife and me, ruling out surgery. *However, he did NOT mention proton beam therapy.*

He recommended Radioactive Seeds (Brachytherapy), which he would do himself. He concluded with the statement that I needed to decide and to come back in a week and tell him of my decision. At this point, I mentioned that there was a new Jacksonville proton beam treatment facility, but he apparently did not know of it, because he mentioned that it must be associated with the Jacksonville Mayo Clinic Facility. It is not; it is the University of Florida, (Shands) Cancer Center. This I did not go into with the doctor. *He did not seem to want to talk any further about protons, so I dropped it.*

Following that visit, I continued to read and study the information available. I found a large number of very good sites on the Internet and several books. One of the most informative sites is Terry Herbert's *"You Are Not Alone"* (YANA) Site; it has everything that a newly diagnosed prostate patient needs to study, including the experiences of men that have had every form of treatment. All prostate cancer patients should join this group.

Another was Don Cooley's *"Prostate Cancer Support Forums,"* which I joined in order to read all of the "feedback" from a great many prostate cancer patients with all different forms of treatment. This was an education! This site has now been converted into a "Ladies Only" forum, and some of the old information has been made available on a new, expanded site: *<http://www.prostate-help.org>*. All prostate cancer patients, newly diagnosed and those further along, should thoroughly explore this site.

Here are some recommendations for new patients gleaned from the original site, and from the new one:

Research and Doctors

1. If your PSA is elevated, and your doctor recommends a biopsy, wait six weeks from the DRE and have your PSA checked again (without a DRE!). Abstain from sex, weight lifting, bicycle riding, horseback riding, and anything that might "exercise" your prostate gland at least one week before the PSA analysis blood is drawn.

2. If you have the biopsy, and it is positive, DO NOT PANIC! Go slow. Resolve to find your own answer for a treatment. Many of us have been down this road before, and there is much to learn before you know enough to choose the solution "best for you."

3. If your PSA is below 10, and your Gleason's Score is 3+3=6 or less), you have time to do all the research you need. With a higher PSA and Gleason 7 or higher, you may need to be more urgent and diligent in deciding on a treatment and taking action, but you still have time to do it.

4. Verify your Gleason number with a recognized expert in the field. Dr. Jonathan Epstein at Johns Hopkins is one such expert. There are other Gleason experts whose names are provided on Mr. Cooley's site [10] and elsewhere.

5. Do NOT under any circumstances accept and go ahead with the first recommendation from the first urologist or specialist! For your sake and that of your family and loved ones, do your homework. Get second opinions. Examine all your options and select the best treatment for you, from the best expert in that particular field that you can find.

6. Get Internet access. If you do not have it at home, your local library will have a terminal that you can use at your leisure, and it will be free if you register. There is a wealth of knowledge and experience there.

I did my own search and phone calls to find out how to have my core slides sent to Dr. Epstein at Johns Hopkins. At this point, I began to realize that my research had "enabled" me, and I was actually becoming an empowered, informed, and educated prostate patient!

As you study and learn, you will come to realize more and more that you alone must make your treatment decision. You can go to six different medical specialists and you will most probably receive a recommendation for the prostate cancer treatment that they each provide. I found the following quote to be very true: *"Prostate cancer has a highly variable natural history, presenting a bewildering variety of treatment management options to the clinician and patient alike, ranging all the way from radical extirpative surgery, to many different forms of radiotherapy, to medical therapy, to no active therapy at all. Each of these treatments has proponents and opponents, and the "perfect" treatment method for this disease remains elusive."* [11]

Eventually, you will find a choice that appeals to you for some reason. The key word here is choice. *You* should make the choice, not the doctor! And the "best" is what *YOU* determine to be the best for *YOU*. The choice should be your decision based on all the facts available to you and your disease and physical characteristics, needs and beliefs. You must be comfortable with your decision.

How do you decide? That is the hard part. You must gather the necessary facts, from the best sources that you can find, study them and do your best to analyze them. You should take your time, and try to think objectively about all possible consequences and your own strengths and weaknesses, and those of your family and loved ones. Whether or not you believe in the power of prayer, thoughtful meditation, or whatever form of quiet reflection that you can engage in, this is the time.

But in the end it should be your decision.

Treatment Alternatives

THE TREATMENT OF CANCER by any means is always a serious and frightening thing, because when diagnosed, the fact of our mortality and susceptibility to life-threatening disease is suddenly thrust upon us. The next step, choosing a treatment method or "modality," as the medical profession calls it, is therefore also difficult. With prostate cancer this is compounded by the fact that there are several confusing choices that neither the medical profession nor the patients can agree on.

Proton beam was my treatment of choice, and much of this book describes that treatment. This chapter describes the other alternatives that I considered, and some of my rationale for rejecting them. Please be aware that my reasons for rejection of a treatment method may not be yours, and that the information here may or may not be applicable to your individual physical condition, financial situation (insurance, work, etc.), outlook on life, or disease characteristics. What I provide is primarily to give a general description of these alternatives, and to show some insight into my decision process. This chapter will be broken up into sub-headings of the other treatments that I considered. Chemotherapy is not discussed because I did not study it and due to complexity. Also, new or unapproved modalities are not described in any detail.

ACTIVE SURVEILLANCE:

"Watchful Waiting" is basically doing nothing. A better approach is more correctly termed "Active Surveillance," and it is indeed a viable treatment method. It is up to the individual patient to decide if this is right for him, based on all available factors about his disease.

If the cancer is very slow growing, and if it is shown by tests to be in the very early stages of development, active surveillance may be a correct approach—*under the watchful care of a good oncologist*—with PSA monitored every two to four months. Some men have adopted this approach with a change in lifestyle and diet, and some have achieved what I would call remission in their cancer, with PSA's either decreasing or holding steady, sometimes for years. Terry Herbert, the author and Webmaster of the "You Are Not Alone" (YANA) site is one of these, and his story should be read.[12] The YANA site is an excellent resource, with many references, and should be a part of your research.

An older patient with other health issues and a relatively slow progression may also opt for the active surveillance approach, because various treatment options may not be viable considering his health, or simply because he chooses this approach to avoid complicating other medical problems that may be more pressing. In these cases, because of the patient's age or the prognosis of some other medical problem, some urologists or oncologists may tell the patient: *"You will probably die of something else other than prostate cancer."*

With younger patients, urologists or oncologists usually recommend an immediate treatment such as surgery, brachytherapy, or radiation. However, the active surveillance approach is just as viable for a younger man as some other modality, especially if the PSA is low, the biopsy diagnosis shows a very low Gleason Score, and there is no evidence of rapid disease progression. It just depends on the individual, and the physical characteristics and/or stage of the disease. A younger man with a slow growing cancer might want to delay even the remote possibility of serious side effects (impotence, etc.), and continue with careful monitoring until there is evidence that the disease is becoming more aggressive.

Some men, however, are not comfortable knowing that the "prostate demon" is alive and in their bodies, and will not accept active surveillance as an alternative. In my case, I had an aggressive, rapidly developing cancer, and active surveillance was not an option. Any patient, young or old, who does not want to take the active surveillance route, should by all means investigate every other option.

SURGERY:

Surgery to remove the cancerous prostate gland is presently considered by many specialists (especially surgeons) to be the "Gold Standard," probably because it has been used the longest of any of the modalities. Many argue that it is the best because it enables the establishment of the true extent of the cancer invading the diseased prostate gland (biopsies within the body only examine a very small portion of the gland, but thorough evaluation is possible with the gland removed). Some surgery also permits the removal of the seminal vesicles or perhaps other cancerous tissue if cancer is suspected to have escaped the prostate capsule. Some also say they would choose surgery because radiation is available if the surgery fails. The flaw in this argument, at least to my mind, is that if the cancer has escaped and is in the pelvic area, radiation will probably be required, either immediately or later as "salvage" treatment. If one is willing to accept the possibility of having to use radiation as salvage, why not use it as the initial primary modality? This is especially true if proton beam radiation (perhaps with adjuvant treatment methods) is considered.

In your study of prostatectomy, you should review the anatomy of the prostate and its proximity to other vital organs. There are many critical steps in this complex operation, and many potential "failure points" (as we termed such points in critical systems when we analyzed a launch vehicle's probability of failure or success).

Blood vessels must be controlled, delicate nerves must be carefully spared, if possible, and vital organs like the bladder and rectum left undamaged. The urethra must be severed in two places, because a section of the urethra is removed with the gland. One of the effects of this may be "penile shortening." The vital bladder external sphincter muscles that control urine flow from the bladder must be preserved during the surgery and when the surgeon reconnects the urethra to the bladder. The greater the surgeon's skill, the lower the rate of incontinence.

Before deciding that surgery is the method of treatment that you want, I suggest that you read several detailed descriptions of the procedure. One is here: <http://www.prostate-cancer.org/education/education.html>. Once at this site, scroll down to "Local Therapies," and Radical Prostatectomy, by Stanley Brosman, MD.

The first nine steps of robotic assisted prostate surgery may be found on the following Internet Web site: <http://www.brighamandwomens.org/urology/Services/Minim ally_InvasiveVideos.aspx> (with video links). Other descriptions may be found in the library or with Internet searches. You should also Google "Prostatectomy Video," and download some of the more exact videos.

Famed Dr. Patrick Walsh of Johns Hopkins says (about the radical prostatectomy):

"The operation is one of the most technically challenging operations in all of surgery.... There are major veins that travel over the prostate that must be controlled, ... There are the delicate nerve bundles, ... and the surgeon must use good judgment about whether these nerves can be preserved or excised to remove all of the tumor, and in preserving them using precise techniques to be sure that they're not damaged. So it's an operation that is challenging, and it's best done by an individual who does a lot of them, who specializes in it." [13]

Make no mistake: prostatectomy is major surgery. One to two liters of blood are usually required to be "banked." These procedures usually take from 1 1/2 to 4 hours. The perineal operation usually takes less time than the retropubic, and may result in less pain afterward. The hospital stay is usually about 3 days, and you will probably be away from work for about 3 to 5 weeks.

Much depends on the skill, expertise, and experience of the surgeon. For surgery, find yourself the very best, one with hundreds of the procedures accomplished.

Surgeon's skill or not, there are profound risks with any complex surgery, and these risks should be completely understood before submitting to the knife.

All of the above applies equally to Da Vinci robotic or laproscopic surgery. There is less blood loss, but the procedure is just as challenging if not more so, and the surgeon's experience and expertise is just as important.

In my case, surgery as an alternative for my prostate cancer was eliminated early on. A surgeon who was honest and forthright informed me that because of my age (then 73) and the fact that I had a stroke in 2003, I was not a candidate for surgery to remove my prostate. I have absolutely no doubt that another surgeon, perhaps not quite so ethical, might have said that I was indeed a candidate. I am afraid that this journey has made me quite cynical regarding some of the medical profession.

When more proton beam facilities become available, the advantages will gradually become a matter of public knowledge. Quality of life factors after prostate cancer treatment will perhaps become the prime criteria for many patients. Perhaps the amazing physics of the proton or other new treatments that are developed or discovered will eventually force the demise of routine radical prostatectomy except in those cases when it can be proven to be medically necessary. I sincerely hope so. In my opinion, the quality of life issues that exist with not only surgery, but also all treatments for prostate cancer, are of prime importance, and should be most carefully considered before reaching your final decision.

HORMONE THERAPY:

Hormone therapy is also called Androgen Deprivation Therapy (ADT) or Androgen Suppression therapy (AST). Hormone therapy can be a valid treatment in its own right or as a supplement in combination with various other modalities. The American Cancer Society's Web site has: *"...The goal is to reduce levels of the male hormones, called androgens, in the body. The main androgens are testosterone and dihydrotestosterone (DHT). Androgens, produced mainly in the testicles, stimulate prostate cancer cells to grow. Lowering androgen levels often makes prostate cancers shrink or grow more slowly. **However, hormone therapy does not cure prostate cancer**.*

Hormone therapy may be used in several situations:

- *If you are not able to have surgery or radiation, or can't be helped by these treatments because the cancer has already spread beyond the prostate gland.*

- *If your cancer remains or comes back after treatment with surgery or radiation therapy.*

- *As an addition to radiation therapy as initial treatment if you are at high risk for cancer recurrence*

- *Before surgery* [or other treatments] *to try and shrink the cancer to make the treatments* [possible or] *more effective"* [14]

Orchiectomy is surgical removal of the testicles, and is considered a form of hormone therapy, since the source of testosterone is removed. This operation is not reversible, and consequently many men reject it. The alternative is ADT (chemical castration) with the use of drugs. One of these drug types is Luteinizing hormone releasing hormone analogs, or LHRH, such as Lupron. There are others, and the proper use of these in the treatment of prostate cancer is best obtained via a physician or oncologist well trained and experienced in treating prostate cancer with them.

The side effects can be quite serious, although many urologists who prescribe the drugs discount most of them, especially when the treatment is "temporary," and given in preparation for some other modality, in order to shrink the prostate gland to a size that can be treated. The side effects can include hot flashes, fatigue, muscle loss, weight gain, irritability, mood swings and emotional episodes, depression, anemia, lowering of "good" cholesterol, breast enlargement and tenderness (gynecomastia), and osteoporosis. I experienced some of these after I was treated with Lupron. Some men may not have the same reactions. There is now a new concern: *"The use of ADT appears to be associated with an increased risk of death from cardiovascular causes in patients undergoing radical prostatectomy for localized prostate cancer."* (From the Web site of the Journal of the National Cancer Institute; Art. # djm168.)

As is the case in most treatments for prostate cancer, controversy is prevalent. From the American Cancer Society Web site:

"There are many issues around hormone therapy that not all doctors agree on, such as the best time to start and stop it and the best way to give it. ... Some doctors think that hormone therapy works better if it is started as soon as possible if the cancer has reached an advanced stage (for example, when it has spread to lymph nodes), if it is large (T3) or has a high Gleason score, or if the PSA starts rising after initial therapy, even though the patient feels well. Studies have shown that hormone treatment may slow down the disease and perhaps even lengthen patient survival. But not all doctors agree.... Some are waiting for more evidence of benefit. They feel that because of the likely side effects and the chance that the cancer could become resistant to therapy sooner, treatment should not be started until symptoms from the disease appear. Studies addressing these questions are now under way."

However, ADT is clearly a useful tool, both alone and along with other treatments for fighting prostate cancer.

CRYOTHERAPY:

Cryotherapy is also sometimes called cryosurgery or cryoablation. Cryotherapy is carried out by inserting (into the prostate) a series of small needles, attached via tubing to a source of cryogenic nitrogen or argon, which super-cools the probe tips. Most cryotherapy units use argon gas at about minus 40 degrees F. The cryoprobe is placed in the proper position within the prostate using imaging guidance, and as internal tissue is being frozen, the physician avoids damaging healthy tissue by viewing the movement of the probe on ultrasound, CT, or MRI devices that transmit images to a monitor.

The procedure for the prostate is complicated, using multiple probes, and requires great skill by the surgeon. In most prostate cancer cryotherapy, an "ice ball" is formed encapsulating the entire gland, and after the appropriate period of time, it is thawed, then the freezing is repeated. Warming fluids are circulated through the urethra to help prevent damage, and there are other special preventative techniques. The recommendations and advice that I received when studying this alternative suggested seeking out an "artist" in the procedure that had done hundreds of them. The main (potential) side effects of cryotherapy are: (1) Usually permanent impotence because the nerves surrounding the prostate are frozen. If "focal" cryosurgery is done so that some of the nerves are spared, this problem may resolve with time; (2) A high probability of urinary problems because the urethra and bladder neck are affected (sometimes overcooled or nearly frozen). This can happen even though there are techniques used to protect the bladder and urethra during the freezing.

Cryotherapy does not seem to be as widely used as some of the other alternatives. One advantage is that it can be repeated. Even though my first urologist recommended it *(or perhaps partly because he did),* I rejected this modality after investigating it.

BRACHYTHERAPY:

"Brachytherapy, also known as sealed source radiotherapy or endocurietherapy, is a form of radiotherapy where a radioctive source is placed inside or next to the area requiring treatment." [15] With prostate cancer the radioactive source is in the form of very small pellets or "seeds," hence the alternative name "seeding." These seeds are usually about the size of a grain of rice although smaller in diameter. The first most commonly used radioactive source is palladium-103, having a half-life (time required to reduce the emitted radiation to one-half the initial level) of seventeen days. The second source commonly used is iodine-125, with half life of about sixty days. Seeds usually have the radioactive source encapsulated in a bio-compatible titanium shell.

As with other modalities, the skill and expertise of the specialist implanting the seeds, and the treatment planning (as to seed location within the prostate) are significant factors. If at all possible go to a "Center of Excellence" hospital that specializes in brachytherapy if this is your choice. Do this if you possibly can, even if it means a long trip and additional time away from home. Additionally, get the doctor that has the most experience in the procedure, not one that has just received his training. Again, this is critical no matter what treatment modality you choose.

"Achieving good results with brachytherapy requires substantial technical skill-...results can vary substantially from one practitioner to another. The dependence on physician's skill is much greater than for external beam radiation therapy. The necessity for such substantial skill has caused some to criticize prostate brachytherapy, because not all patients will achieve the same results as those treated by physicians who specialize in prostate brachytherapy. The implant process is the summation of the physician's skill and the expertise of the team, which

includes a radiation oncologist, urologist, and radiation physicist. Clearly, radioactive seed implants are a highly operator dependent procedure, which can have a steep learning curve." [16]

The choice of the radioactive source to be used may also be important, because studies have shown a higher complication rate with the iodine-125 than with palladium-103. One study showed a grade 1-2 complication rate of 8% and grade 3-4 complication rate of 7% with Iodine-125 versus 3% and 1% respectively for the palladium-103.[17] A newer development is the use of Cesium-131 which has a half-life of only 9.3 days. *"Cesium-131 is the most significant scientific advancement in prostate cancer brachytherapy in more than 20 years."* [18] However, long-term results are not yet available for the use of Cesium-131 in prostate therapy.

There is also the fairly new High Dose Rate (HDR) brachytherapy, in which higher dosage "seeds" are inserted temporarily to provide the radiation. With HDR "temporary" brachytherapy, plastic catheters are carefully inserted into the prostate gland, and then a series of radiation treatments are provided through these catheters. Using computer controls, the highly radioactive source (Iridium-192) is inserted into the prostate through the catheters, targeting the locations of cancerous tumors, and avoiding areas near the urethra and rectal wall as much as possible. Of critical importance are CT scans, the imaging during treatment, and careful treatment planning. The computer controls the dosage by timing the duration that the individual seeds remain within the prostate. When the treatment is complete, the plastic catheters are removed, and no radioactive seeds remain in the prostate gland, and this is the advantage. However, the radiation dosage is higher and thus the potential for damage may also be greater. The issue of seed migration (see next paragraph), however, is eliminated with HDR seed treatment.

Another factor with standard brachytherapy was the issue of "seed migration." As I mentioned in another section of the book, some sources indicate that this is a low incidence problem, and when it does occur, there are few degrading consequences. I researched the subject of seed migration, and found instances of some severe consequences, and a fairly high rate of incidence. The primary concern of specialists like Dr. John Blasko was the reduction of radioactive dosage to the cancerous targets if seeds were lost soon after treatment, which studies showed usually to be the case. What happens is that the tiny seeds somehow "come loose" from the prostate tissue into which they have been implanted, and "migrate" to other parts of the body through blood vessels, usually to the thorax, where the heart and lungs reside. One study was done using chest X-rays to locate and identify seeds that had left the prostate after brachytherapy. *"The results of the study found that 55% of the 100 patients had one or more seeds identified on the CXR."* [19] Another study was done to review whether or not "linked seeds" that had been specially prepared and "stranded" together, prevented individual seed migration.

"The use of linked seeds embedded in vicryl sutures for the peripheral portion of permanent radioactive seed prostate implants significantly reduced the incidence of pulmonary seed embolization in patients treated with the Seattle technique." [20] Apparently this stranding does help.

Brachytherapy results certainly seemed promising in so far as "cure rates" for low risk PCa. Finally, I located a study that compared the results of proton beam therapy directly with brachytherapy, using case-matched data.

The "cure" results were similar, however the conclusions were: *"High dose EBRT* [proton beam therapy] *is equivalent to brachytherapy for control of localized prostate cancer. Ultimately, treatment decisions may be based upon **future quality of life comparisons**."* [21]

RADIATION:

Radiation is a general category that includes not only X-ray *photon* treatment, but also *protons*. Brachytherapy is also classified as radiation by most radiologists (it is actually internal radiation), but in this book I provide the description under the previous separate heading.

Radiation or radiotherapy accomplishes its purpose by damaging the genetic material (DNA) within the cancer cells, which prevents future cell multiplication and growth of the cancer. The actual damage to the material using radiation is called ionization, which requires a high level of energy to be effective. With external beam devices (which is what this heading deals with), complex and expensive equipment is needed in order to deliver this high-energy radiation.

There are many different types of radiation, such as heat and light from the sun, microwaves that we use in our ovens, and radio waves used for communication. Here we are only concerned with radiation that can be purposefully elevated to a sufficient energy level to cause the wanted DNA damage to a cancerous tumor residing in the human body. The specific forms of radiation primarily used for this purpose are X-rays and Gamma rays (*photons*) and atomic particles (*protons* in this book). [22]

When radiation is used to treat cancer, it is usually provided over a period of time, in what are described as fractionated doses, or a series of treatments that provide a total dosage of radiation prescribed by a treatment plan. The treatment plan is usually designed by a radiologist or radiation oncologist in concert with other specialists, even including medical physicists, after obtaining test results that completely describe the characteristics of the disease and physical condition of the patient. These tests will usually include, in the case of prostate cancer, the patient's PSA, the Gleason Score from biopsy results, bone scans, CT Scans, MRIs, and other diagnostic tools and formulae to characterize the state of the disease, and other tests and study of medical records to establish the patient's physical condition.

The total prescribed dosage for a particular radiation treatment plan is usually measured in *grays*, abbreviated Gy. The *gray* is a unit of measurement for the absorbed dose of radiation, as defined by Louis Harold Gray in 1940. One gray is the absorption of one joule of radiation energy by one kilogram of matter. The dose for each treatment is calculated by dividing the total prescribed treatment dosage by the number of planned treatments.

There are established protocols for the allowable total dosage to be provided in any radiation treatment that should be followed by the radiologist when preparing a treatment plan. For example, the protocol at Loma Linda University Medical Center when I received my proton treatment was a total of 79.2 Gy, administered in forty-four treatments, once per day, five times per week, at 1.8 Gy per treatment. This may now have been increased.

When used in this manner, the treatment is called "radiotherapy," as opposed to a single treatment session (akin to surgery), when the total prescribed dosage is provided in a single application. By using smaller individual doses instead of a single dose, damage to healthy surrounding tissue is reduced, and additionally, the intervening time allows any adjacent tissues that received radiation to begin a healing process.

Photon radiation used in most X-ray radiotherapy is produced by a machine called a photon linear accelerator. The devices that produce this photon radiation are relatively inexpensive *(compared to the systems required to produce protons)*, and are in use in many hospitals or medical facilities worldwide.

There is a "special case" of photon radiation used that does not require a linear accelerator. This radiation is produced by radioactive elements such as cobalt-60 and cesium-137, and is called Gamma ray radiation. The machines that are used are extremely heavy (due to shielding) and expensive, and there are not very many hospitals that have them. *These machines are not used to treat prostate cancer.* They are mentioned here simply to describe another form of photon radiation used for some cancer treatment. Gamma ray treatment of cancer is currently limited to head and brain tumors, and in some cases neck and cervical spine. The "Gamma Knife" is the trade name for one such machine, which works by a process called *stereotactic* radiosurgery, and uses multiple beams of radiation (usually 201 radioactive sources) converging in three dimensions to focus precisely on a small volume, such as a brain tumor, permitting intense doses of radiation to be delivered at one application. The critical nature of the target requires the patient's head to be restrained in a frame while the treatment is given.

Gamma rays and X-rays differ in their origin. Gamma rays originate in the nucleus of an atom, while X-rays originate in the electron fields surrounding the nucleus or are machine-produced, as they are in the linear accelerator. *But they are both photon-based radiation.*

Photon treatment (X-ray), or conformal radio therapy (3D-CRT), of cancerous tumors using linear accelerators is by far the most prevalent radiation method used for prostate and other cancers, with several variations. Most discussions and written material found on the subject of radiotherapy refer to this photon-based treatment. There are more recently developed systems such as intensity modulated radiation therapy (IMRT), image-guided IMRT (IG-IMRT), and the so-called "CyberKnife" technology. With CyberKnife (sometimes termed CK), damage to surrounding healthy tissue or organs is reduced by manipulation of the radiation beams and additional "compensators." This is accomplished by targeting the tumor to be treated and using a computer controlled robotic mechanism to aim the radiation from various angles. This is again termed "stereotactic" application. Treatment that is usually fractionated over several days may be referred to as stereotactic *radiotherapy* as opposed to stereotactic *radiosurgery,* as is used with the Gamma ray treatment of tumors, and sometimes with CyberKnife (the CK treatments are usually provided in one to five applications). The CyberKnife machines are not as expensive as the Gamma machines, and are becoming more prevalent. Additionally, they can be used for a wide variety of human cancer types and locations, *including prostate cancer.* According to the CyberKnife Society Web site, (*<www.cksociety.org>*) there are currently (2007) seventy-three CK locations in the United States, and thirty-eight at locations in other countries.

The side effects of *photon* radiation therapy for prostate cancer are sometimes quite serious, but other times less so. The reaction caused in normal body tissue through which the radiation passes on the way to the tumor being treated and again coming from it varies from person to person. The type of radiation treatment and prescribed dosage has an impact on the body response. Some potential side effects are:

- Fatigue. During treatment this is common, but can sometimes be mitigated by exercise.
- Skin. Redness, tenderness, dryness and peeling. This generally goes away with time.
- Nausea and diarrhea. Once common with standard X-ray (photon) treatment, this reaction has diminished. The possibility does still exist because of possible irradiation of the bowel area.
- Bowel problems. There may be fistulae, rectal bleeding and/or permanent change in bowel function.

- Urinary. Because the prostate gland is in such close proximity to the bladder, and the fact that the urethra passes through the prostate itself, there may be urinary incontinence, bladder problems, plus painful and frequent urination.

- Sexual impotence. Erectile dysfunction (ED) can develop almost immediately or up to two years following the completion of the radiation. It may be permanent. The radiation may affect the nerves surrounding the prostate that control erection.

The most serious of these side effects are avoided to a great degree with PROTON radiation. **Proton** producing machines require a completely higher level of complexity (the machines are based on nuclear physics), and are particle accelerators known as cyclotrons or synchrotrons. These systems strip off protons from the parent atom, typically hydrogen, and accelerate them to the required therapeutic energy level, usually on the order of 60 to 300 million electron volts. A later chapter (Chapter Nine) provides a complete discussion of proton beam radiation therapy (PBRT).

Most discussions and written material currently found on the subject of radiotherapy refer to ***photon***-based treatment. It is seldom that ***protons*** are mentioned or discussed, probably because until recently there were only three centers in the United States that could provide the treatment. The following is a good example: *"Radiation therapy is the treatment of cancer with ionizing radiation. Radiation works by damaging the DNA (genetic material) within the tumor cells, making them unable to divide and grow. Radiation is often given with the intent of destroying the tumor and curing the disease .... However, although radiation is directed at the tumor,* ***it is inevitable that the normal, non-cancerous tissues surrounding the tumor will also be affected by the radiation and therefore damaged*** *(Burnet N.G. et al. 1996).* ***The goal of radiation therapy is to maximize the dose*** *to tumor cells while minimizing exposure to normal, healthy cells (Emami B. et al. 1991)."* [23] [Bold Emphasis added.]

The preceding paragraph is addressing **photon** radiotherapy, because it says: *"it is inevitable that the normal, non-cancerous tissues surrounding the tumor will also be affected by the radiation and therefore damaged."* This acknowledges the differences and the primary disadvantage of photons compared to protons.

Another key sentence above is: *"The goal of radiation therapy is to maximize the dose to tumor cells while minimizing exposure to normal, healthy cells ...."*

In July of 2005, a comparative study was done by Ulricke Mock et al. of the Department of Radiology, University of Vienna, in Austria, to determine if there was any advantage to *proton* beam treatment of prostate cancer compared to conformal photon therapy *(3D-CRT)* or photon based intensity modulated radiation therapy *(IMRT)*. The results indicated the proton advantage:

"With both photon techniques non-target tissue volumes were irradiated to higher doses (mean dose difference [greater than] 70%) compared to proton-beam radiotherapy. Differences occurred mainly at the low and medium dose levels, whereas in high dose levels similar values were obtained. In comparison to conformal 3-D treatments, IMRT reduced doses to organs at risk (OARs) in the medium dose range, especially for the rectal wall.
Conclusion: IMRT enabled dose reductions to OARs in the medium dose range compared to 3-D conformal radiotherapy. ***A rather simple two-field proton-based treatment technique*** *[Alternate left and right "through the hips" treatments, one per day for prescribed total dosage, as is done at LLUMC.]* ***further reduced doses to OARs compared to photon-beam radiotherapy.*** *The advantageous dose distribution of proton-beam therapy for prostate cancer may result in* ***reduced side effects,*** *which needs to be confirmed in clinical studies."* [24]

Thus the conclusion that IMRT is less damaging to surrounding tissue and organs than normal conformal photon radiation was demonstrated, *as well as the conclusion that proton beam radiation therapy is less damaging to surrounding tissue and vital organs than either photon based system, both 3-D CRT and IMRT.*

The experience at LLUMC during the last fifteen years has substantiated this research study, and future clinical studies by the new proton facilities around the world will further validate the conclusion.

The primary problems with **photon** radiation are that without manipulation, the energy will be deposited near the surface of the receiving body, and also that the energy continues on through the body and in so doing, harms normal healthy tissue and organs in its path. This occurs (to a lesser degree) even with newer, more sophisticated **photon** systems such as IMRT, and this "collateral" damage is basically why **proton** therapy was developed. Because of the physical properties of protons (explained in Chapter Nine), damage to nearby normal healthy tissue and organs is minimized if not completely eliminated.

There are circumstances that may rule out the use of protons for some. If a man does not live near a Proton Center or cannot travel to one; or has a demanding job so that he cannot take the eight or nine weeks off to have the treatment (or cannot arrange to do his work remotely via computer); or has no insurance coverage; he may have to find another alternative. If he *IS* near a Proton Center, a schedule *CAN* be arranged; the daily treatment time is short, and most proton centers operate two shifts. There may be other reasons that preclude proton treatment. For these, the patient can certainly choose another treatment modality that will fit his circumstances better.

In my case, I had reviewed my options quite thoroughly prior to discovering protons, and factors affecting quality of life had become primary criteria in my decision process. When I found proton therapy, it was easy to realize that protons should be on my "short" list.

Decision!

FINALLY, I DISCOVERED the last key in my decision making process. I was reading a very informative page: *http://www.prostateproton.com/*, by a former LLUMC proton patient, and found a reference to a support group: *"... It's very important to read these **testimonies** from many of these men at **Proton BOB** to help you make an informed decision about your own Prostate treatment."* [25] I entered "Proton BOB" in my search engine. The very first result was the "Proton BOB" Homepage. I navigated to the page, and this fascinating site took about five hours of my immediate time! I strongly urge anyone with cancer to read the testimonials provided on this Web site. The entire site is most informative and educational, but the testimonials are a very important part of the proton story. While there are successes with other treatments, I challenge anyone to find such an extensive listing of glowing testimonials for radical prostatectomy or *any* of the other modalities.

Robert J. Marckini, a prostate cancer survivor who received his LLUMC proton treatment about seven years ago, authors this site, and is personally responsible for all the interviews that resulted in the formation of the "Brotherhood of the Balloon." I found the following: *"Numerous studies have shown that all prostate cancer treatment options cure cancer at statistically the same rate. Even Johns Hopkins, the 'Prostate Cancer Surgery Capital of the World,' in a 2005 Health Letter has stated, 'Recent studies demonstrate that newer types of radiation are as effective as surgery, which was once thought to be the surest way to a cure. And as radiation techniques grow ever more precise, side effects may be more limited and treatment and recovery made easier.'"* [26]

I now realize that this could be an oblique recognition of proton beam therapy!

As I said, I spent about five hours on this site, and became more excited as I read. ***The treatment and results seemed almost too good to be true!*** (Have you heard or thought that before?)

I had two questions that did not seem to be answered in the information that I found. I found Bob Marckini's email address, and sent him the following message:

In a message dated 10/17/2006 4:57:52 P.M. Eastern Daylight Time, cnsjones@bellsouth.net writes:

As a newly diagnosed prostate cancer patient, I write to you to ask one or two questions that do not seem to be addressed in the information or testimonials that I have read on the Protonbob site or elsewhere. Perhaps you will have some knowledge or insight that will help me to understand. I am wondering about radiation damage to the urethra and its passageway through the center of the prostate. No one seems to have any problems associated with the PB Therapy, and yet I read many instances where the standard radiation treatment of prostate cancer causes many urinary problems.

If the energy delivered to the prostate is so confined and intense, how is the urethra in the capsule relatively unharmed?

This is really the only concern that I have, other than the "balloon" procedure, which it seems to me would press the rectal wall right up against the prostate area to be treated. What am I missing here?

Any thoughts that you could share will be most appreciated.

I hope soon to be a member 0f BoB! I am corresponding now with Loma Linda and the new facility in Jacksonville. If you hear of any reports from the JAX treatment center, please let me know. I talked with them today, and to date they have only treated 12 patients, which is somewhat of a concern."

Decision!

Bob Marckini's almost immediate response:

"Dear Fuller,

Regarding the urethra, I had the exact same question six years ago. What I've been told, and what I have seen thousands of times, is that the urethra seems to be surprisingly resistant to radiation damage. While it's in the "line of fire," it weathers the radiation well. I do not know of anyone who has had problems with damage to the urethra [from Proton Therapy].

The balloon performs two important functions: 1) it pushes the prostate up against the pelvic bone, helping to immobilize it, and 2) It "inflates" the rectum, thus preventing radiation damage to most of the circumference of the rectum.

Regarding Shands vs. Loma Linda, please give me a call at my home in Massachusetts. I'll be happy to talk with you about this.

--Bob Marckini"

I called Bob, and he most graciously and factually answered my questions. Bob told me that he was writing a book, *"You Can Beat Prostate Cancer...,"* and we discussed writing in general.

As a genealogist, I had published a family genealogy on my family. In one of the "BOB Tales" newsletters, Mr. Marckini mentioned that he was looking for a publisher of his book, and I responded with an email that I had used an Internet "Print-On-Demand" publisher called Lulu.com. I got an immediate response from Bob. In a remarkable coincidence, that was where he had decided to first publish his book!

I purchased his first "Pre-Release" copy. This book was one of the final determinants in making my decision for a treatment of my cancer. Bob Marckini has now released the final edition of his book. The complete title is: *"You Can Beat Prostate Cancer, and You Don't Need Surgery to Do It."*

After correspondence and conversations with four more Loma Linda Proton "Alumni," one a medical doctor, and another a radiation oncologist, my decision was made! I would seek proton beam therapy to treat my cancer.

Why was this? In my case it ended up with the simple question: "Which treatment is most likely to provide the *Quality of Life* that I want for the rest of my years?"

I had no desire to risk being incontinent, becoming suddenly impotent, having surgical or freezing damage to vital organs, risking photon radiation damage, or having the possibility of "migrating" radioactive seeds.

For me, after finding and understanding proton beam radiation therapy, it became almost a no-brainer. Just exactly as Bob Marckini had expressed it, I felt as if a great weight was lifted! It was as if I had been guided to this solution, and suddenly my path was clear! This was one of those few moments in life when somehow you just *know* that you have made a correct decision!

I knew that I would continue to investigate Loma Linda and the other proton centers, but I somehow *knew* that my "Proton Decision" was the correct one.

Everyone that I spoke with or corresponded with who had proton treatment there urged me to consider Loma Linda. There seemed to be some almost mystical aura about the hospital! This was most difficult to understand at the time. I was a patient in different hospitals before with several surgeries, including the prestigious George Washington University Hospital in Washington, D. C., and they were all about the same, seemingly only interested in making sure that you had insurance coverage, and getting you in and out as quickly as possible. How could LLUMC be any different?

Now, having experienced the caring and dedicated way that all patients at LLUMC are treated, I understand. The missing element that I did not even think of was the fact that this hospital is a religious institution, and is dedicated to the complete treatment of patients: physically, mentally, emotionally, and spiritually. The motto of the hospital is to *"Make Man Whole."* This they do with a continuing effort, and the entire staff participates, from the person who picks up the used gowns in the dressing rooms to the highest administrative official!

However, at that time I had no clue about any of this, and I was more immediately interested in the new University of Florida Shands Proton facility at Jacksonville, since it was only about three hours from our home. I tried several times for three days to call the number that I found on the Internet, and kept getting a recording: *"Leave a message and I will return your call."* I even called a Shands physician's number "blind" in hopes that they would know how to reach the Admissions Department of the new center; they did not. Finally, after several days I did make contact, and the person promised to send me all the forms that I needed and an information packet. (Note: This lack of timely response now appears to have been corrected.)

While waiting on Jacksonville's new Proton Center to get back to me, I also contacted the new Proton Center at M. D. Anderson Hospital in Houston, Texas. I had decided that I did not want to go to Massachusetts General in Boston (also I do not particularly care for big cities) or to Indiana University Hospital at Bloomington in the winter, so did not contact them. The Admissions people at M. D. Anderson said they would send me a package, however after I told them that my Gleason's Score was 4+4=8, they told me that I would not be accepted because they were only accepting new patients with a Gleason of 7 or below! At first, the only reason that I could fathom for this was that they wanted their initial success rate to be as high as possible, and so were not accepting "higher-risk" patients. But now I realize that perhaps they prioritized prostate cancer patients in order to provide treatment for other cancers. There are more prostate patients than patients with other forms of cancer. Still, higher-risk patients are accepted at some other centers, and treatment protocols varied as needed.

For me, that left the Florida Proton Institute at Jacksonville or Loma Linda University Medical Center in California as my possible treatment locations. I was still convinced at this point that I should stay in Florida, close to our home of forty years, for my cancer treatment. I did some more checking on the Jacksonville location, and discovered that there were very few rentals available near the new facility (this may now have changed). I would probably have to find a place twenty or thirty minutes away from the hospital. This was not a terrible drawback, but did enter into my decision. When I found that the treatment protocol was for forty-four or forty-five treatments, one per day, five days a week, I knew that at either Jacksonville or Loma Linda, I would have to have temporary housing near the location or within a short driving distance. And again the "big city" was a factor.

I went to the LLUMC Web site again, and there were many contact numbers and every possible description! I called the 1-800-PROTONS (1-800-776-8667) number, (another is 1-800-496-4966) and was connected with a most knowledgeable person, who asked all the questions about my cancer and reports, then gave me the fax number to send them. She then said that I would be receiving their "New Patient" Package the next day! They Fed-Ex'd it overnight, and I did! I was starting to become more impressed with this Loma Linda outfit! In all of my searches about LLUMC, I found only good things. Also, the Web site has now been vastly improved, and has videos that should be viewed. See: *http://www.protons.com* and navigate to the videos.

Decision!

In addition, I found that on the date that I initially contacted the Jacksonville Shands Proton Center, they completed treatment of their *twelfth* patient! LLUMC had started (pioneered) proton beam cancer therapy in 1990, and had completed over *eleven thousand!* And more than sixty-five percent were prostate cancer patients! Thus, on October 20, 2006, exactly one month after I "discovered" "The Beam," my final decision was made! I sent my records to LLUMC for review by Dr. Carl J. Rossi.

Previously, I had also sent a fax to Dianon Labs, (who prepared my original biopsy report) asking them to send my slides to Dr. Jonathan Epstein at Johns Hopkins for a second opinion reading. This famous doctor is considered by many to be the foremost authority in interpretation of prostate cancer biopsy samples. I had found the recommendation to do this (to get a second opinion on your Gleason's Score) on a very informative Web site that I had found and studied extensively.

October 24, 2006 was my follow up with my new urologist. This was a most surprising and interesting meeting, although very brief. I sat in his office, and asked a few questions regarding the Lupron that he answered. Then I told him that I had decided to try for proton beam therapy at Loma Linda in California. His immediate response, no hesitation, was: ***"Good for you!"*** I had prepared some data on PBT and Loma Linda, and gave it to him. He thanked me for the information, reached over and shook my hand, and said to be sure to have all results and records sent to him. On the way out of the office, he put his hand on my shoulder and said: ***"Enjoy your stay in Loma Linda, it is very beautiful; I have been there!***

I was totally shocked, but looking back, I realize now that he obviously knew of Loma Linda and proton beam therapy; he no doubt had been there for some reason, perhaps a symposium.

This illustrates and emphasizes the fact that a patient *MUST* do his own research, find the alternatives, and reach his own decisions. As in that Web site that I found early on, this was a good example of *"What your doctor may not know and [or] may not be telling you."*

Wednesday, October 25, 2006: I received the phone call from Loma Linda University Medical Center. My records had been received, Dr. Carl J. Rossi had reviewed them, and I was accepted for a consultation. We made the appointment for 9 a.m. on December 26, 2006, almost exactly four months since my diagnosis. The treatment was to commence in early January 2007. They Fed-Ex'd another package with housing, a tape, and other up-to-date information, which we received October 26. On Wednesday I also received a call from the LLUMC Insurance Department: Medicare would pay most of the cost, and my Blue Cross/Blue shield would take care of the rest. There would be NO out of pocket cost for the proton therapy! I would of course have the expense of the trip and housing, but some of this would be recovered if I itemized on my income tax return.

Thursday and Friday, October 26 and 27: We consulted the housing list that Loma Linda had sent, and selected a small house to rent. We corresponded with the owners via E-mail, and they even sent us pictures of the interior and furnishings! We called and confirmed it. It was to be available by the time we arrived in California; the patient living there would be just finishing up his treatments then. The house was just four blocks from LLUMC; I could *WALK* to my treatments! Later I found that the area surrounding the hospital was filled with short-term rentals, which had developed as the need arose during the early days of proton therapy. When their treatment was completed, the patients returned home and the next patient moved in, usually within a day or so.

LLUMC maintains a list of these homes and apartments for the newly arriving patients. (Most centers do this.) On my street alone there were seven other prostate cancer patients, and my back-yard neighbor was also one. This was another strange coincidence. This neighbor was Gregg Moore, and it turned out that I had been corresponding with his wife Laurel about proton beam therapy for about a month before we left for California!

Tuesday, October 31: I Received the report of the second opinion review of my biopsy slides from Dr. Epstein of Johns Hopkins. The highest GS was 7 (NOT 8), but was the more aggressive 4+3=7. However, there was a notation: "One sample had evidence of Perineural Invasion." I read what I could of this, and sent Dr. Rossi a message asking him to clarify what this might mean in terms of my treatment. It turned out not to affect anything after further examination and MRI tests showed that the cancer was still confined to the prostate capsule. In addition, the conformal proton beam would encompass a margin of about ten millimeters outside of the total gland that would treat any such invasion.

All of the above was a remarkable series of events regarding our plans and preparations that just seemed to "fall in to place." Dr. Rossi had an opening the day after Christmas; the rental was available the day after our planned arrival in California; and my family's pre-Christmas annual get-together in Alabama fit into the plans perfectly. It was as if the entire sequence was pre-planned, and this reinforced my feelings that God had a hand in leading me on this journey! I now know that this was true. Prayers work! The further along on this journey I went, the more I knew it.

My sister later had a similar experience, after facing surgery to removing a cyst that had grown into her spine, and was causing intolerable pain and a near inability to walk. Her first doctor had told her that the cyst must be removed and two vertebrae would have to be "fused" together. We were all concerned that this might cause future problems. She had the operation scheduled, but the prayers of our entire family intervened, and a nephew arranged for a consultation with a neurosurgeon who was normally so busy that there was a two-month waiting time for an appointment. Once again, things just seemed to fall into place, like they were pre-planned! After a few tests, this young but very experienced specialist informed my sister that he could simply remove the cyst through a small opening; that there should be no difficulties involving the spinal cord; and no need for any work on the vertebrae. She would go home from the hospital the same or following day.

There was a providential cancellation in his surgery schedule, and he accomplished the procedure within a few days. My sister's pain was immediately gone and she had no further difficulties! Prayers work, and second opinions not only are good, they should be mandatory!

My wife and I started to plan our trip "out west," and included a four-day stop in Alabama to see relatives, just before Christmas. Our plan was to arrive in Loma Linda on December 19, 2006, after a trip of over 2,600 miles. I now know that I was not the first to call this Loma Linda journey a "Radiation Vacation," but that is what we started calling it and that is what it turned out to be!

While there, we became most relaxed, caught up on our reading, and I also started writing this book. It was a most relaxing and stress-free time, and this was most unusual since we had traveled thousands of miles to a strange place for a major medical procedure, in the hope of successfully treating a life-threatening disease!

Chapter Six

Loma Linda and "The Hospital"

WE ARRIVED IN LOMA LINDA on December 18, 2006, spent one night in the Loma Linda Inn, about one block from our rental house, and moved in on December 19, 2006. What a delightful surprise this "home-away-from-home" was! Two bedrooms, newly refurbished and completely furnished, it even had a one-car garage. One bedroom was set up as an office, with high-speed Internet! I had brought my home computer and the next day I was up and running. I would continue in my normal routine with writing and research, and was able to attend to all those things that we had committed to do "on-line."

That morning, I went in to the hospital to check things out, and the first impression was that "This place is huge!" The main hospital, which has about 900 beds, is a major trauma center that handles accident victims from all of the surrounding area. There is a helicopter landing-pad on the top of one of the main buildings to handle trauma patients. I looked up and saw one helicopter that had just landed, blades still rotating, and another was already lifting off! To the left of the imposing main hospital building is the Children's Hospital, considered to be one of the very best in the world. I later found out that the very first infant heart transplant had been done there. And beneath the Children's Hospital was the Proton Center!

We went in to the hospital again on the 20th about nine o'clock, and "lucked in" to a tour, that lasted about three hours. In addition to information about the treatments and the hospital, there was a follow-up presentation detailing the many benefits provided by the hospital for the patients during their nine or ten week stay in the Loma Linda area. I was totally saturated!

In our group of new patients, there was a man from Australia and his companion (as he introduced her), a Chinese man who now lives in California and his wife, a man from Arizona who was trying to decide on treatment, a man from Connecticut and his wife, and ourselves.

There was also a lady from Chicago who was being treated for breast cancer as a part of a new pilot program. Dr.'s Bush and Slater of LLUMC have published a report on this, and it is already drawing interest and comment in the radiation oncology world. [27]

Her story was compelling, to say the least. She had resigned herself to the dreaded standard radiation treatment at a prestigious Chicago hospital, and was actually driving on her way to get her first treatment, when she heard on her car radio about the new Loma Linda program for proton treatment of breast cancer. She actually turned around and went back home to investigate, canceling her appointment. She was accepted into the treatment program at Loma Linda and was about halfway through her treatment regimen.

The Proton Radiation Center is located in a massive underground three-story concrete and steel structure, which forms the foundation of the Children's Hospital. Concrete walls up to fifteen feet thick and plate steel and concrete ceilings accomplish radiation shielding. The structure was also designed to withstand earthquakes! Across the street toward the west, is a three story parking garage for the hospital staff that is the size of a large city block. One must keep in mind that Loma Linda University Medical Center is not only a hospital, but also a large medical and dental university, one of the most respected in the country. All in all, the entire hospital and university campus is one of the most impressive complexes that I have ever seen. I am surprised that the hospital is not more widely recognized by the public, but many doctors and members of the medical community nationwide definitely are aware of the hospital and its reputation. When I mentioned my plans to receive proton beam therapy at LLUMC to my new G. P. in Florida, she was aware of Loma Linda's expertise in children's heart transplants, and had heard of the Proton Center.

Loma Linda is without doubt a college town. It is not a large city, and this was a factor in my selecting Loma Linda for my treatment. This one institution, Loma Linda University and the associated Medical Center, overlooks the entire small city. Not only is the hospital and university the primary employer in the city, it dominates the landscape as well. The university and faculty buildings and facilities abound, but are almost eclipsed by the huge and imposing structures comprising the hospital. The hospital is Seventh-day Adventist, and Saturday is the local Sabbath, so most businesses observe it. Mail is delivered locally on Sunday instead of Saturday.

Because LLUMC is a Seventh-day Adventist facility, it "makes no apologies" for being first and foremost a religious institution. This fact seems to have influenced everyone there. The staff and in fact all employees seem to have been infused with a helpful, caring attitude! From the very first day, when we walked in the front entrance, every staff member that we came in contact with went out of his or her way to be helpful and courteous.

The hospital is huge, and a stranger can certainly easily get "lost" in the many hallways. We did not get lost, but would have without the directions given by a kind individual who saw our indecision at a hallway corner and stopped to ask if he could help, and did!

We were not "strangers" at all. Patients, including "out-patients" (which is what proton beam therapy patients are) are considered "Guests" of the hospital. All patients and their "caregivers" are provided official badges identifying them as such.

Since we arrived in Loma Linda and checked in with the Loma Linda University Medical Center early on the 19th of December, I was officially a proton patient, and as such both Carolyn and I were invited to a very special event that occurs every Christmas. Dr. J. Lynn Martell, the Special Assistant to the President of LLUMC, invites all proton patients who are in Loma Linda, away from home and family, to dinner on Christmas Day at his home. This has become a tradition for the last seven years, and continues. The number of patients who have attended this event has varied from forty-five to eighty-six. This year about seventy signed up, but *only* sixty-seven attended, including us. The fellowship and food were wonderful, and Dr. Martell and his wife Karen, as host and hostess, were very gracious! This was yet another example of the caring attitude of this wonderful hospital and staff. As proton patients, we were not alone in a strange place, without family, for the Christmas holiday! Dr. Martell tells the story of how the idea for this suddenly came to him during one of the regular Wednesday night Patient "Support" meetings (another story), and he made this invitation to about fifty attendees. Over forty accepted the invitation, and *THEN* when he got home, he told his wife he had invited a *FEW* guests for Christmas dinner! Remarkably, as he tells it, she remained married to him!

Both my wife and I were becoming more and more aware of why the "Loma Linda Experience" is almost legendary! Certainly almost every member of the hospital staff that we encountered was extremely caring, friendly, helpful, and efficient.

The hospital has even dealt with the parking problem (the parking lots are congested to say the least), and continues to do so with additional long-range plans. Not only do they have trams to pick you up from your parking place and take you to the hospital door, they also have valet parking. In addition, you simply have to call or ask at the reception desk, and a shuttle bus will come and pick you up and take you to or from any of the hospital's remote locations or offices, usually within five or ten minutes. The shuttles will also take you to or from such places as the Loma Linda Inn, which was just down the street from our "home away from home" on Yardley Place. So if I did not feel like walking to or from my treatments, or if it rained, I could ride!

We met with Dr. Carl J. Rossi for my initial consult on the day after Christmas, December 26, 2006. I was very impressed with his professionalism, but Dr. Rossi was even more helpful and clear in his explanations than I had hoped. He spent all the time required to thoroughly go over all aspects of the treatments, then took the time to answer each of the questions that I had prepared before the consultation, even though some of them had been already explained. Dr. Rossi's Case Manager, Nurse Sharon Hoyle, was also most helpful with the preliminaries and getting us "oriented."

It turned out that Dr. Rossi and I had a little more in common. He had done some undergraduate work at the Kennedy Space Center in 1987, and was a runner as I was back at that time. There are many trails and routes to run around the many canals and roads to tracking sites on the KSC lands, and during lunch time usually several astronauts, and other KSC employees would go out for a five or six mile run. Dr Rossi did this also while he was working out of the "O & C" Building (Operations and Checkout). My office at the time was also in the O&C, and I routinely went for a run three times a week. The uncommon thing we had in common was the fact that a wild hog had chased us both! I actually got between a mother pig and her piglets (this is as bad as getting between Mama Bear and her cubs)! Luckily there was a side trail that I took at flank speed! Dr. Rossi said that a wild boar popped out of the palmettos just ahead, and he did a quick reverse and sprinted for half a mile! I am sure that we must have seen each other during these runs, and waved in passing! All this developed because I had remarked that he (the doctor) was wearing my brand of running shoes!

Later in this consult, I found that I was to receive a special MRI the following day to rule out any spread of the cancer to my pelvic area. Then in the afternoon I would be fitted to my "Pod" (I will explain this later) and perhaps get a date set for my first treatment.

At this point, I was full of anticipation, but in no way worried! I just simply *knew* that I was in the best place and would receive the best treatment in the world! My experiences during the entire time that I was a patient at Loma Linda University Medical Center enhanced and confirmed this early impression.

Once the proton therapy was started, and with the realization that only about an hour each day would be required for the treatments, suddenly everything became very relaxed! There were no pressures, and my dear wife really enjoyed the fact that all of the housekeeping duties that normally occupied her time at home were at a minimum, as was cooking, since we both enjoyed eating out. We became experts on the local restaurants, and this may have contributed to my weight gain that occurred in spite of exercise at the Drayson Center, that we did three times a week. We took long walks in the neighborhoods surrounding the hospital, enjoying the hills and views of the distant mountains. We went on day trips to Lake Arrowhead and Big Bear, and toured the NASA Dryden Space Center at Edwards Air Force Base. We both really had a good time at our "home away from home," and the total experience really was a "Radiation Vacation!"

Preparation

THE LONG AWAITED DAY ARRIVED when I was to get my "Pod" made and the required CT Scan done. In my mind, these were the first steps in my treatment, since they had to be done before the proton beam treatments could begin. In addition, the scheduled MRI was to be done in the morning. In an attempt to start out in an organized fashion, I prepared my schedule for the day as follows:

Wednesday Dec. 27 2006:

Schedule:

8:30 - Go in to Center – Take paperwork for Lupron. Pay for MRI, Get copy of orders to take with me.

9:45 - MRI at Off Center Site; Professional Plaza

11:00 - Grab some lunch – Cafeteria?

12:00 - Void, then Drink 16 oz water and leave for CT/Pod appointment at 1230, B Level.

1:30: - Clinic A Level Get Appointment times and assignment. - Home around 2:00?

5:00 - Wednesday Night BOB Meeting

The above plan did not work out. It started out wrong, and continued that way all day because the first problem cascaded causing scheduled times to get changed. However, during this trying day I remembered some of the words from *Desiderata:*

"Nurture strength of spirit to shield you in sudden misfortune. But do not distress yourself with dark imaginings..." and persevered!

At 8:30 a.m. I was in the Proton area; I gave the paperwork for the Lupron to Melissa at the Proton "A Level" Desk, to give to my Caseworker. I asked for a copy of my MRI Order to take with me. Melissa spoke with someone, who said that the MRI Office at the Professional Plaza already had the orders.

I took the shuttle, got there at 9 a.m. for my 9:45 appointment, and then found that they in fact did NOT have my orders, so I should have insisted.

LESSON LEARNED: ALWAYS TAKE A COPY OF ORDERS FOR TEST WORK AT A REMOTE SITE!

Why the MRI? My initial consult with Dr. Rossi included a discussion of whether there was any possibility that the cancer had spread beyond the prostate gland. In Florida, I had both a CT Scan and Bone Scan, and the results were negative for spread of the disease.

However, cancer cells are microscopic, and there is always the possibility that some indication was missed in the evaluation of the resulting images. Dr. Rossi had mentioned that if the disease had spread, then both proton and photon (regular radiation) treatments should be used; the proton to destroy the cancer in the prostate gland, and the photon to take care of any disease that had spread beyond the prostate capsule to the pelvic region. Naturally, I hoped for only proton beam therapy because of the minimal side effects.

But I wanted to be sure that we were doing the job right! It would have been very easy to simply say: "Lets go with the protons only." But there was that chance of spread, and Dr. Rossi suggested that an MRI could help make that decision. I agreed and it was scheduled. This is a good example of the cooperation and communication between the specialist and an educated, informed patient that results in treatment decisions that are most likely to be the best that can be made.

Preparation

What is MRI? *"Magnetic Resonance Imaging (MRI), formerly referred to as Magnetic Resonance Tomography (MRT) or, in chemistry, Nuclear Magnetic Resonance (NMR), is a non-invasive method used to render images of the inside of an object. It is ... used in medical imaging to demonstrate pathological or other physiological alterations of living tissues."* [28] This permits the doctor and radiologist to look at the surrounding organs and tissues in the pelvic region for additional signs that the cancer had "escaped" the prostate capsule. One type of MRI, used by Loma Linda physicians and radiologists in advanced diagnosis of prostate cancer, is called "Prostate MR Spectroscopy" and is special because (in my case) in addition to the MR scanner, a CP array abdominal coil and a Medrad endorectal coil were used.

Having arrived at the "Professional Plaza" off-site MRI Center at nine o'clock, I sat and waited for about an hour while they got my MRI order from the Center, which was finally faxed over.

In the meantime another person "took" my appointment time. Finally at 10:55 they started my MRI, which took about 50 minutes. This was a "special" MRI, and required the insertion of a "coil" in my rectum next to the prostate gland; this was much more uncomfortable than a DRE, but not as bad as the biopsy. When this was complete, it was a little after noon.

Since my appointment for my Pod and CT scan was for 12:30 at the Medical Center, I did not eat lunch, and rushed to get back to The Proton Center "B Level" for my appointment for my Pod and CT. I arrived at B Level at 12:25, checked in, and drank my water (16 ounces). The water, along with the balloon, serves to stabilize the prostate. The water also distends the bladder and moves most of it out of the proton path during treatment.

How relieved I was that I had made my appointment time! But I need not have rushed. About 1:30 p.m. "Tony" came for me, and took me down the hall to get my pod made. On the way he told me that there might be a short delay for the CT scan, but that if I could wait, we would get it done. I said "Sure!" I was anxious to get everything moving in the right direction!

Why was the CT necessary? The definition for CT found on the Internet: *"Computed tomography (CT), originally known as computed axial tomography (CAT or CT scan) and body section roentgenography, is a medical imaging method employing tomography where digital geometry processing is used to generate a three dimensional image of the internals of an object from a large series of two-dimensional X-ray images taken around a single axis of rotation."* [29]

In other words, the CT machine is an extremely sophisticated computer controlled device that takes X-ray "slices" of a particular part of your body. The computer can then manipulate these images to allow the radiologist to determine many things, including the three dimensional shape of the target, and the relative location of various bone structures and organs; in this case my prostate gland's position in relationship to bones and organs in the pelvic area. So the CT Scan is used for establishing the definitive position location of the organ being treated for cancer by the proton beam.

For prostate cancer patients, the location and position of the prostate gland with the initial CT scan (which is necessary to be done only once) is used in preliminary positioning for all follow-on proton treatments and for other special planning operations, including making the special apertures and the "bolus" device used to "conform" the proton beam three-dimensionally to the size and shape of the prostate being treated.

What is the reason for the "Pod?" The pod is simply an immobilization device. Anyone who has ever had a CT (or an MRI) is aware that the instructions are always to lie very still and do not move while these marvelous machines do their work. "The Beam" as I call the proton beam throughout this book, is controlled so precisely that any movement of the patient must be minimized. The very imaginative people at LLUMC came up with the idea of holding a person's body in the same position for all the treatments by "restraining" them in an individual "Pod," made to conform to that person alone. This is done through the use of a section of "PVC" pipe about eighteen inches in diameter and about seven feet long, cut lengthwise in half, and closed off at the ends. The patient lies down in his individual pod (the bottom half of the pipe) on a sheet of plastic, and a mold of his body is made using a foam mixture.

The pod procedure begins with entering a dressing room and putting on a standard hospital gown. All clothing is removed except your socks. (I was embarrassed since I was wearing "holy" ones and my big toe was protruding!) Then your attendant takes you down the hall to the Pod Room. The pod, with fixed attach points, is lying horizontally next to an access platform with steps, and you then lie down in it on top of a large plastic sheet, feet firmly against the bottom half section of the pod closure. At this time, I was informed that I was about to be initiated with the "Secret Handshake!" I was instructed to roll over on my right side, which is rather difficult to do while in the pod, but I managed it. Then a lubricated balloon is inserted into your rectum, and filled with warm water (approximately four ounces). This is to position and hold your prostate gland securely for the CT Scan, and the same procedure is done for each proton treatment. This "initiation" is the procedure that led to the formation of the "Brotherhood of the Balloon," or BOB, that will be described and explained later in this book.

For the proton treatment the water-filled balloon serves to secure the prostate against the pelvic bones for the proton treatment, and also to move most of the delicate rectal tissues away from the radiation of the protons. Drinking water before treatment does the same for the bladder. The balloon procedure was not uncomfortable, no more than a DRE. After this was done, I resumed my position in the pod. A mixture of expanding foam was poured into my pod underneath the plastic sheet (like the "Great Stuff ™" foam sealants used for home insulating and sealing projects). The two chemical components of the mixture are stirred up in a large "milk-shake mixer" and then poured into the pod underneath the plastic sheet. The mixture starts to "foam" and rises up around your body, conforming to your unique shape. Thus your pod is yours and yours alone, and serves to immobilize you in the exact same position when you are brought into the Treatment Room for your proton beam treatment, and also for this initial CT Scan.

I can truthfully say that this period of time, about thirty minutes or so, while I was in the pod and the foam was rising and expanding underneath me from the waist down, was the most relaxing time that I spent on this particular day. As the foam expanded and rose up around the lower half of my body. It felt very warm, but not uncomfortably so. I actually drifted off to sleep for a few minutes!

When the foam had hardened, the balloon was removed, and I had to exit the pod and go back to the dressing room to get dressed, to await my turn on the CT machine. Then I went back to the Proton Reception Room to wait some more. Ordinarily, I would have gone right in for the CT. The initial delay of the morning at the MRI facility continued to cause problems with my schedule for the rest of my day!

Preparation

When Tony came out to get me for the CT Scan my nerves were once again starting to get the best of me. I had been awake since 5:30 a.m. with nothing to eat but crackers from the snack bar since the night before. Not knowing when I was going to be called for the CT Scan, I did not want to chance leaving the waiting room and missing the last thing standing in the way of getting my first appointment with The Beam!

Once again I went to the dressing room, got prepared, and then was placed in my VERY OWN pod. Again I received the balloon. Then I was wheeled into the CT Room, and carefully positioned for the scan. This painless "feel-nothing" procedure took about twenty-five minutes. I was then taken back to the Pod Access Room and exited the pod. After I again got my clothes on, I was reminded to go up to the A Level desk to get the date and time for my first proton treatment. I got to the desk at 4:05 p.m., and the receptionist had left to go home five minutes before! The end of a long and tiring day had finally arrived, and my day was complete with one last late-again event! I would not be able to get my appointment time until Thursday morning.

I went to the front entrance, caught the Shuttle Bus to the Loma Linda Inn, and at 4:35 p.m. I was back at my Yardley Place "home away from home," and was warmly greeted by my anxious wife.

We proceeded to get ready to go to the Wednesday night Proton Patient Meeting. That meeting, which we were to attend regularly for the entire duration of my treatments, was the high point of my day. And, there were vegetarian snacks and juice provided! The warmth and fellowship of my companion cancer patients served to remove all remnants of the unfortunate delays of the day, and completely restore my feelings of faith. I knew that I was in the right place, and that I was to receive the very best treatment in the world for my cancer!

One new patient stood up and gave his story: He was a Seventh-day Adventist and so knew of Loma Linda, but knew nothing of the proton treatment. In doing his research he found that Loma Linda University Medical Center had the proton beam, and after studying the history of proton treatment, his decision was made! He is an engineer with the Boeing Company at Huntsville, Alabama, and was working on an important project with a deadline. He just could not take off the time that would be required for the treatments. The Boeing Company was remarkably supportive of his treatment decision. They set him up in an apartment in Loma Linda with all new computer equipment necessary for him to continue working on the project at the remote location.

This, you see, is the thing about proton beam therapy: There is no overnight hospitalization; you are an outpatient, and can continue all your normal daily activities. Only an hour and a half per treatment day is required to be allotted, to allow for schedule variations. The treatment itself, after the required set-ups and careful verifications that take twenty minutes or so, only takes about two or three minutes!

Therefore if it is at all possible for a person to work "remotely" using the computer, or if commuting is a possibility, then a schedule can usually be worked out that will facilitate the proton therapy sessions. This should be considered no matter which proton center is your choice.

The following day (Thursday, December 28) I called to get my first Beam treatment appointment time. It was to be at 3:00 p.m. Thursday, January 4, 2007.

Later, while I was running errands in the afternoon, Nurse Sharon called at my home phone. She said that the CT Scan must be redone, because the balloon was apparently not inserted or filled correctly the day before!

I had forgotten my cell phone, and my wife could not reach me. So when I arrived home, I found that it was too late to get the scan redone that day.

Since this was New Years Weekend, with the hospital closed on Saturday and Monday (New Years Day), the earliest that I could get my CT Scan redone was Tuesday, January 2, 2007. So once again I was back in the waiting game! Now my worry was that this would cause my first treatment date and time to be delayed!

Thinking back, I remembered that the second balloon did not quite feel the same as the first one, not as "full." Being unfamiliar with the procedure, I did not recognize that something might be wrong.

LESSON LEARNED: IF SOMETHING DOES NOT FEEL OR APPEAR "RIGHT," LET SOMEONE KNOW!

On January 2nd, I was at the B Level desk at 8:a.m. After telling Levita, the receptionist, what had transpired about the previous CT Scan, she went off to find out what was going on. In a few minutes, Tony himself came out, and told me that they were planning to get me in as soon as possible, but that an emergency had come up and that they were standing by to receive a child who was under anesthesia. He suggested that I come in at 11:45 and they would get the scan done.

So I called my wife; she came and met me, and we went for a two-mile walk. Right after we walked in the door when we got home, the phone rang! The CT Team had found that the emergency case was going to be delayed, and they could get me in right away!

I "quick-timed" it back to the Center, and Tony was ready and waiting. I quickly got into my gown and was once again introduced to my pod and the balloon!

Tony took extra precautions with the balloon, taping it into position. This really went fast! It was about 10:30 a.m. I was again wheeled down the hall to the CT Room, met the attending radiologists, and positioned for the scan. This time everything went like clockwork! The CT scan was completely successful. I was "sliced" 98 times, with the computer-controlled machine moving two millimeters between each picture. The scan took around twenty-five minutes, and I was done!

By 11:30 I was back "home," and all of the preparations that I was required to participate in were complete. The Proton Staff now had to complete my treatment plan, including manufacturing my own unique metal aperture and wax "bolus" that shape the Beam to fit my prostate gland, using the data from the CT Scan. The unique "Modulator Wheel" that "spreads" and fits the Bragg Peak of the proton beam to my prostate had to be configured also. According to information that I had received, this process can take as long as about two weeks. In my case, this time only was eight days from the consult, so some of these preparations had been ongoing.

The next day was Wednesday, January 3, 2007, and we had an appointment that morning with Dr. Rossi to discuss the results of the special MRI and get his recommendations. This was very brief and to the point. He told us that there were no indications on any of the scan results that the cancer had escaped the capsule, and that I was cleared for a proton only treatment. I believe now that this was primarily because I was already on the Lupron hormone therapy, and that if I had not been, I would probably have had a combination of protons and photons recommended. My Gleason Score was 4+3=7, and this is classified as intermediate grade prostate cancer, and more aggressive. As it was, he recommended a continuation of Lupron during the treatment.

Prostate Cancer Meets The Beam

MY FEARS THAT THE DELAY because of the repeated CT Scan would also delay the start of my proton treatment were unfounded! The appointment to introduce my cancerous prostate to The Beam was on for 2:30 p.m. January 4, 2007.

On Wednesday, January 3, after the meeting with Dr. Rossi, we accomplished our first exercise session at the "Drayson Center," for one hour. This is a complete facility, two stories, including every imaginable machine, a complete weight room, an indoor 1/10-mile exercise track on the second floor, a basketball court, five handball courts, and an Olympic size outdoor swimming pool. Oh, also a hot tub. Some past proton patient had donated two expensive treadmills (in addition to the eight they already had) for use of proton patients only. The use of this facility is *free to all patients and their "caregivers."* It is heavily used by the faculty and students, which is why the treadmills were donated—the (wealthy) donor got tired of having to sign in and wait for a treadmill! This Drayson Center is another "world-class" addition to the LLUMC facilities.

The next day, January 4, my wife and I arrived at the hospital about a half hour early and checked in with Levita, the receptionist on Level B. We were both sore from the exercise the day before. We had not really worked out since three days before we left Merritt Island (we belong to the "Island Fitness" center there—They extended our membership for the duration of my treatment. Very nice of them!)

My first treatment was scheduled for 2:30 p.m. There were about six or so patients waiting for their daily

"Dose" of protons. This number is about normal, since there are three gantry treatment rooms and one horizontal beam line (HBL) room used for treatment. Sometimes there are a few guys there just for the fellowship and camaraderie. Many "hang out" before or after their treatment to talk with other friends and patients, or to use one of the two computers available (they have internet access).

Scheduling times are not precise, because some set-ups take longer than others. In addition, sometimes children under anesthesia must be treated, and they always have priority. This time a tech came in and said that they had a child due in under anesthesia, and that there would be a delay, so that some rescheduling would be necessary. This of course had already happened to me once, although I found later that it was a pretty rare occurrence. He asked for volunteers to reschedule, and all of those waiting volunteered. Some of us asked to be put on a "call list." We were told that we could leave if we wanted, and they would contact us when normal scheduling resumed. I opted to leave, and my wife and I went for a long walk, then home to wait.

Apparently the catch-up process took awhile, because it was not until 8:00 p.m. that a technician at the hospital called, and said if I could come in, they could go ahead with my first treatment! *YES!!*

I called the University Shuttle Service and asked for a pick-up at the Loma Linda Inn, where we had stayed the first night when we arrived, and was right at the end of our street. I quickly walked down to the corner, and five minutes later, I was picked up and whisked right over to the hospital. I went immediately to B-Level and to the dressing room to get ready for my treatment. The changing room is just down the hall from the HBL room, which was my assigned treatment area.

Prostate Cancer Meets The Beam

Through "the luck of the draw" I had been assigned to the horizontal beam line treatment room. The only difference between the HBL and the "Gantry Rooms" is that the beam nozzle in HBL is stationary. As with the gantry rooms, the patient (in his pod) is carefully aligned with the beam nozzle, and the prostate alignment verified (within a millimeter) with X-rays and comparison with the CT Scan information. This preparation takes about 15-25 minutes.

As soon as I was in my hospital gown, (open at the back), with nothing else on but my socks and a tee shirt (and watch), Randy, the tech, came and escorted me to the treatment room and my pod. The massive fixed beam nozzle dominated the room, and off to the side was the smaller "Eye Beam" nozzle. I was assisted into my pod, which was already attached to an alignment platform and cradle with extremely precise vertical and horizontal adjustment capabilities. After I was properly positioned in the pod with the balloon inserted, the initial preliminary treatment positioning was accomplished. Incidentally, all feelings of modesty are very quickly lost after the first couple of times of having the balloon inserted, even by team members of the opposite sex!

Also, my unique conforming aperture and bolus were inserted into the beam nozzle mechanism, and the special modulator wheel inserted into the rotation mechanism just behind the nozzle. Then bar codes affixed to them were scanned, along with my pod and other setups, into the computer activation system. The scanned bar codes unique to each patient must all match, or the computer-controlled process cannot start, and the treatment will not proceed. The proton beam nozzle is about five feet off the floor, and so my pod/cradle and I were raised about another two feet, so that my hips were in close proximity to the Beam nozzle. At that point the X-ray positioning scans are accomplished so that the prostate and surrounding bony structures can be aligned to the primary CT scans that were done during the preparatory process.

This is a careful adjustment, done incrementally by horizontal and vertical movements of the pod/cradle, and is done with an accuracy of millimeters (using a digital readout display) until the CT scan and X-rays "match" for the prostate location and other primary references.

In the HBL, it is necessary to develop the film X-ray pictures to verify and determine alignment adjustments that are required. This takes several minutes.

As I lay there I became aware of a deep muted humming sound, almost a vibration or roar that was just there in the background, overriding all other noises. I found out later that this was the great synchrotron and supporting power supplies and equipment in the adjacent facility, which was separated from the treatment room by about fifteen feet of reinforced concrete wall.

I had been in my pod for all this time during these preparations (about twenty minutes), and the pod is quite comfortable, so I was almost asleep when, as the team left, they told me "Here we go!" After about a minute, I heard a high-pitched sound like a loud fan whirring. This was the "modulator wheel" spinning up to about 300 revolutions per minute. This is the device used to "spread" the Bragg Peak, allowing the Beam energy to focus and conform to the entire prostate. I then heard a humming sound that lasted for about one or two minutes. There were also "beeps" from a Geiger counter, (so I was told) but my hearing aids did not pick up their frequency. I lay there wondering when the treatment would start, when the team came back in and one of them said: "That's it! Come back at 8:30 in the morning for your second treatment!"

I had just been "Zapped" with 225 million electron volts of protons and felt absolutely nothing; it was almost a letdown!

The balloon was removed, and I was assisted out of the pod, got dressed, and went up to the front entrance to catch the Shuttle Bus home. My wife and I watched TV for a while and went to bed. It was "One down, 43 to go," with another scheduled for Friday morning!

Friday, January 5, 2007: I got to Level-B at 8:00 a.m. and went directly to the dressing room. While there, there was an announcement over the P. A. System for a moment of prayer, that I found very comforting! Then the day team (there are two shifts) welcomed me, I went through the same routine as the night before, and at 9:10 a.m. I was getting dressed to go home!

About 10:00 we went to the Drayson Center for exercise; this time a little over an hour, then went home and had a pleasant relaxing Friday! Again, I thought to myself: "Man, this Proton Therapy is TOUGH!"

The treatments were all similar, but I will describe one more treatment day from my notes, a Wednesday, which includes my regular weekly "See the Doctor" day.

"Wednesday Jan 10, 2007: Treatment #5 at 0630. Walked both ways. The treatment was on time."

Later my wife and I walked back over for the weekly consultation with the doctor. This is a required once a week consultation to discuss the treatments and any problems. I had none except that Dr. Rossi wanted me to continue the Lupron (hormone) therapy until the proton treatment was over. I had hoped for this to be finished, so that I could start the recovery process from the Lupron. I had been told the recovery period was just as long if not longer than the total duration of the Lupron injection (four months). The objective of the Lupron in my case was to shrink my prostate gland, and to shut down the body's process of making testosterone because the prostate cancer "feeds" on it.

The side effects for short-term hormone therapy or ADT (Androgen Deprivation Therapy, in my case Lupron injections) initially were not terribly serious, but later they did bother me. In addition to the loss of sexual function (it is chemical castration), I had hot flashes, mood swings, and experienced loss in muscle strength. I also gained some weight, about ten pounds, in spite of exercise.

I put off the scheduled next injection but finally decided to go ahead, at least for one more month. I rescheduled for the following week. It seems that there is a treatment protocol history that indicates six months of ADT yields slightly improved statistics for outcome, especially if the last two months are during treatment. I later chose to have just this one additional one-month injection. I wanted to start recovering from the Lupron.

As my treatments progressed, I lost the sense of being a patient, and my treatments became simply a part of my daily activities. I actually looked forward to going in for my daily dose of protons, because once that was completed, the entire day was available for many of the varied activities available in this part of Southern California. I had specifically requested a treatment time of 7:00 to 7:30 a.m., and after a week or so of being bumped around at various treatment times, this assignment was granted and my treatments became pretty regular, and even with some delays, I was usually walking on my way "home" by 8:30 in the morning. When Friday came, this was even better, since the entire weekend was free! There are no treatments on Saturday or Sunday. Plus, by the time I arrived "home" there was usually a nice breakfast awaiting, along with a cup of coffee. I know, I should NOT have been drinking coffee, but I have just not been able to give it up!

I really liked the early morning schedule. I usually left for the short walk to the hospital about seven a.m., and it was a very quiet and peaceful time. There were not many cars out on the streets near the parking lot yet, and this enabled me to really enjoy the early morning light and the distant mountain views. It was winter, and the temperature was usually in the mid forties, cool and crisp. Unbelievably, one morning I stepped out the door onto snow! It was the first snow in seventeen years! Later on in January and February, I had to be careful not to slip in icy patches on the way. This early morning time became my "quiet time" and a time for reflection about how lucky I was to have found protons and Loma Linda.

Description of the Treatment (B-Level) and Patient Reception (A-Level) Waiting Areas:

These areas are simply great, with many things to make a patient comfortable. On the B-Level, these are: Two computers with Internet access and a printer; a large bookshelf with all kinds of books, fiction and non-fiction, that patients can take home and read; a big table and several LARGE picture puzzles for puzzle fanatics; a large salt water aquarium, very nicely populated; two TV sets and comfortable chairs and sofas for watching; a separate alcove for children, with toys, a child size table and chairs and a small TV set up for watching children's DVD's. There is also a counter set-up with cold and hot water, with tea and hot chocolate and usually some kind of snacks. All this and a most accommodating receptionist who seemed to know everything!! This is just on the lower B-Level, two levels underground, where the synchrotron and supporting equipment is located, and the treatments are given. Some patients even come to the waiting room on B-Level to socialize and take advantage of the perks of the facility. There is even a piano!

A few of these things are also available one story above (the A-Level) where the main reception area is for new proton patients and where the patients see their doctors for the initial consultation and weekly visits.

The day before the last of my forty-four treatments, I asked for and received the "first slot" in the daily HBL Treatment sequence, at 6:30 a.m. When the final "ZAP" of protons was done, I had a brief good-by session with my caring crew, took some photos of them and the proton equipment, and presented them with souvenir Kennedy Space Center NASA caps. It was a sad-happy experience, knowing that this was probably the last time that I would see them, but I was *READY* to return home, in spite of the impending 2,600-mile drive. We had been almost three months away from Florida.

The previous day (a Wednesday) I went back to the hospital after breakfast for the obligatory "check-out" visit with the doctor. As luck would have it, Doctor Rossi was out of town, and it was necessary for me to meet with one of the other radiation oncologists on the LLUMC Staff. Thanks to Nurse Sharon Hoyle, Dr. Rossi's Case Manager, this turned out to be Dr. Ben Rodney Jabola.

This turned out to be a great "exit interview!" Dr. Jabola spent over an hour with me, going through my complete record and treatment books page by page. He led me through the complete treatment preparation and planning sequence, explaining all the radiographs and other planning documents. He carefully explained why the proton treatments were the preferred method instead of normal X-ray radiation, drew diagrams, and actually showed me on my very own charts and diagnostic radiographs. He went "above and beyond" what was necessary in doing this, and I was not even his regular assigned patient!

I was once again astounded at the caring and helpful attitude exhibited by all the members of the Loma Linda University Medical Center staff and personnel. Dr. Jabola is one of the best! But then, I did not meet all the other "best" members of the Loma Linda staff!

Chapter Nine

Protons

WHY PROTONS? As mentioned earlier, it is because of the *elegant physics of the proton particle*. Proton beam radiation therapy or PBRT is classified as external beam radiation therapy (radiotherapy), because it targets the tumor with an external source of radiation energy. But this radiation is not the same as X-ray radiation; it is actually a beam of sub-atomic particles called protons.

Proton beam therapy is classified as radiation therapy, however I believe that it should be considered a special form of radiation. In my opinion that there is a sufficient difference in the physical characteristics of standard photon or X-ray radiation and proton radiation to give protons a separate classification. Perhaps this may eventually happen as more proton centers become active.

The facts regarding the physics of the proton particle drove the early research into the medical possibilities of proton therapy. In 1946, brilliant scientist Robert Wilson authored a paper entitled *"Radiological Use of Fast Protons"* that became the basis for all the follow-on work of modern proton radiation therapy. Robert Wilson and his team built the world's first high-energy proton accelerator (cyclotron) at the Fermi National Accelerator Laboratory in Batavia, New York. Later work led to the first use in cancer treatment at the Berkeley Radiation Laboratory about 1954, although this was actually an experiment in a non-hospital environment.

The Harvard Cyclotron was later used to treat several cancer patients, again in the laboratory environment. This early work led to the development of the first use in a purely hospital environment at Loma Linda University Medical Center, with a synchrotron designed specifically for proton beam radiation therapy.

The driving force behind the development of this facility was Dr. James M. Slater, another brilliant and respected physician, and one dedicated to the Loma Linda philosophy of providing the absolute best in treatment and patient care, delivered in a helpful, caring and healthy environment. This overall theme and philosophy emerges time and again in the history of LLUMC.

The reader needs to clearly understand that ***protons are NOT photons, and they do NOT react with human tissue in the same way as photons.*** There is a basic physical difference, based on the physics of atomic structure. Protons are particles, like tiny, guided pellets.

There are many who do not understand that X-ray radiation treatment is ***photon*** treatment. All "standard" radiation, even the advanced intensity modulated radiation therapy (IMRT), and "CyberKnife" radiation is ***photon*** based. Photons are like tiny "waves," and tend to "scatter" in human tissue and continue through the body.

Protons are different. I like to say that protons are "like a rifle compared to a photon shotgun!" I have read another apt comparison: Bob Marckini says, *"Like a smart bomb compared to a cluster bomb! There is much less "collateral damage."* Professor Kia Iwamoto, of the Radiation Oncology Department at UCLA says: *"Radiation Therapy is all about collateral damage! We want* [at least] *95% of prescribed dose to the entire tumor, and less than 5% to the surrounding critical structures."* Particle beam radiation therapy (in this case the proton beam) does this best because of the physical properties of protons and what is called the Bragg Peak.

Several years ago, Dr. Carl J. Rossi of LLUMC provided an update on proton beam therapy. This paper (written about 1995), while now slightly out of date, provides an overview by a well-recognized specialist in the field of proton beam therapy. As such, it is important, and is provided here:

Protons

Conformal Proton Beam Radiotherapy of Prostate Cancer

Carl J. Rossi, Jr. MD; Asst. Prof., Radiation Medicine LLUMC

"Radiation therapy has long been an effective and accepted treatment for localized prostate cancer. The efficacy of radiation therapy is directly influenced by the radiation dose, which is delivered to the prostate. Conversely, treatment related acute and late toxicity is dependent on the dose of radiation given to the surrounding normal structures and the volume of these structures, which receive high radiation dose [photons]. **Conformal proton beam radiotherapy takes advantage of the low entrance dose, sharply defined high dose region ("Bragg peak"), and lack of exit dose inherent in the proton soft tissue interaction to deliver precision high dose radiotherapy to the prostate while simultaneously limiting radiation dose to the bladder and rectum.**

The use of proton beams in clinical radiation therapy was originally proposed by Dr. Robert Wilson in 1946. The specific application of this modality to prostate cancer treatment began in the mid 1970's at the Massachusetts General Hospital via the efforts of Dr. William Shipley and associates. The world's first hospital based proton beam treatment facility became operational at Loma Linda University Medical Center in the fall of 1990.

Pre-treatment Planning

In October 1991 a program for the treatment of organ confined prostate cancer with conformal proton beam radiotherapy was initiated. All participants in this program were subject to the usual pre-treatment evaluation (pre-treatment PSA, pathology review, bone scan).

For conformal proton beam planning purposes the patients were immobilized in a custom-fitting plastic cylinder. A balloon was placed in the rectum and inflated with 120 cc of water so as to exclude the majority of the rectal volume from the prostate treatment fields. A thin slice CT scan of the pelvis was performed with the patient in his immobilization cylinder.

The prostate, bladder, and rectum were outlined by a physician on LLUMC's 3D conformal planning system, which utilizes beam's eye view planning and dose-volume histograms

93

to optimize individual treatment plans. In virtually all instances a simple two-field (right and left lateral beams) plan provided for the best coverage of the gland while sparing the majority of the bladder and rectum. All individual cerrobend apertures and wax tissue compensators were produced on automated milling machines controlled by the 3D planning system.

Patient position was verified daily before each conformal proton beam treatment by obtaining orthogonal radiographs of the patient's pelvis in the treatment position via a coaxially mounted X-ray tube. Measurements were made from various bony landmarks to the isocenter and compared with optimal measurements obtained from a computer-generated digitally reconstructed radiograph (generated from the CT planning scan) and moving the treatment table accordingly. Typically, the time ... for each patient's set-up and treatment was 20-30 minutes per day.

Treatment Protocol

Between January 1992 and December 1995, 260 patients with early stage prostate cancer (defined as stages T1-T2B, PSA < 15, NO or Nx, no prior hormonal therapy or surgery) were treated. The radiation dose to the prostate was 74-75 CGE [Gy] (Cobalt Gray Equivalent, utilizing a proton RBE of 1.1) given at a dose rate of 1.82 Gy per day. Two hundred and nine patients were treated with conformal protons alone while fifty-one were treated with a combination of 30 CGE protons to the prostate and seminal vesicles plus 45 GY conformal photons to the pelvis.

Treatment Results

Median follow-up is twenty-six months, with a range of 3-56 months. Biochemical failure was assessed using the published definition of Zietman (three successive > [greater than] 10% elevations in PSA, or a single dramatic PSA elevation which provoked initiation of androgen suppression). Late radiation morbidity was scored using the RTOG morbidity scoring system. The Kaplan-Meier method and Log-Rank test were utilized as necessary for statistical analysis. The actuarial four-year freedom from biochemical relapse for all patients in this early stage group was eighty four percent.

When broken down by clinical stage, the chance of showing biochemically no evidence of disease (bNED) was 97 % for stage T I patients, 83 % for stage T2A, and 80% for stage T2 B (p = 0.27). Patients with Gleason's score 2 [to] 6 tumors had an 83 % chance of being bNED, while Gleason 8 [to] 10 patients had a 68 % chance of being bNED. All patients with pre-treatment PSA's of < [less than] 4.1 were bNED at 4 years vs. 82% with PSA's > 4.1 [but] < 10.0 and 80% for PSA's >10.1 [but] < 15.1 (P 0.29).

As has been demonstrated in other published reports we found that the PSA nadir is a strong predictor of ultimate biochemical outcome. Analysis of patients in whom a minimum of twenty-four month follow up had been obtained demonstrated a 95% actuarial 4 year bNED in patients whose nadirs of ...[equal to or less than] 0.51 ng/ml vs. 78% for nadirs [greater than] 0.51 [but less than] 1.0 and 40% for nadirs [greater than] 1.0 (P<0.0001).

[P-value: A statistics term. A measure of probability that a difference between groups during an experiment happened by chance; i.e. a p-value of .01 (p = .01) means there is a 1 in 100 chance the result occurred by chance. The lower the p-value, the more likely it is that the difference between groups was caused by treatment].

Complications

The most prevalent late treatment-related morbidity is asyptomatic and/or minimally symptomatic rectal bleeding. For reporting purposes we assigned a late morbidity score of two to any patient who on rigorous questioning mentioned any rectal bleeding whatsoever. If a patient required two or more blood transfusions this would have been assigned a score of three, while bleeding requiring major surgical intervention (i.e., colostomy) would have been scored a four.

At four years, the actuarial incidence of any rectal bleeding is 40% **but to date none of our patients have experienced any grade 3 or greater rectal toxicity.** *Fifteen percent of the patients have experienced a grade two GU toxicity (intermittent hematuria, increased urinary frequency) but* **none of the patients have developed a grade three or greater GU morbidity.** *"* [30]

[End Article. Bold emphasis added.]

The following is my attempt to describe the reason that protons are used for cancer treatment. For additional information or more technical data on the physics of protons, simply do a search for "Proton Radiation."

One definition of a proton is: *"A stable, positively charged subatomic particle in the baryon family having a mass 1,836 times that of the electron"* [31] This is the start of understanding why the proton is better suited to radiotherapy than the standard X-ray (photon) treatment. One significant fact here is the mass of the proton compared to electrons (photons); the protons are (comparatively) very large and heavy particles which will penetrate matter—in the case of radiation medicine this means human or animal tissue—only to particular depths, based on the kinetic energy of the beam. In addition, protons are charged particles. This enables them to be manipulated by means of magnetic fields. This means that they can be separated from the nucleus of the atom that they are associated with, (in this case, hydrogen) their velocity (energy) increased or decreased, "focused" into a narrow beam, and accelerated and "guided" (within high vacuum devices) by means of extremely powerful magnetic fields. At LLUMC, the device that accelerates the hydrogen protons to about two-thirds of the speed of light is a form of cyclotron known as a synchrotron. The energy of the beam can approach 250 million electron volts (MeV), but the usual energy level at LLUMC for prostate cancer is about 225 MeV.

The most important characteristic of the proton used in radiation medicine is due to what is called the "Bragg Peak." Due to the remarkable physics of the proton particle, when it reaches the depth in the human body that has been targeted, it essentially stops. When it does so it suddenly releases the kinetic energy stored in the particle, and this ionizing energy is particularly damaging to the reproductive ability of cancerous cells.

Protons

The precise location of the prostate is verified before **each** treatment. By knowing the exact location of the target, the energy of the beam is adjusted such that the peak energy is deposited in the prostate, and is also spread out to envelope the entire cancerous target. As the beam exits the powerful equipment it is "shaped" three dimensionally with special apertures and machinery to conform exactly to the prostate. In fact, a "margin" of about ten to twelve millimeters beyond the gland is treated, and this provides assurance that any cancerous cells in the process of "escaping" are treated. A "Modulator Wheel" designed for the target size and shape is used to "spread" and reduce the sharp Bragg peak.

Almost the same procedure is used for other cancers, and the precision of the beam can be controlled sufficiently to treat very small brain tumors and even tumors of the eye. For brain or eye tumors, special "Head Masks" conforming to the individual's face and head are used to position the patient in the same location for each treatment and to assure that there is no movement during the radiation.

Conversely, photons (electrons or X-rays) have no mass, scatter easily when they impact human tissue, and continue on their way until they exit the body, causing damage to not only the cancerous target, but also normal tissues and organs as they pass into and through them on the way to the target, then again on the way out from the target. Modern methods have been devised to attempt to minimize the damage to surrounding delicate tissues, such as conformal IMRT (intensity modulated radiation therapy), where the target is attacked from different angles with varying intensities, but none of these methods are equivalent to the proton beam in so far as total radiation (the integral dose) of non-cancerous tissue when used with the proper treatment planning and the well developed systems, machinery, and equipment.

Chapter Nine

Dr. James Metz, editor of "Oncolink," which is provided by the University of Pennsylvania and is a tremendous resource to use for cancer treatment research, provides a great explanation of protons and a further discussion of the biologic effect differences between protons and X-rays. They are basically the same. (Note: This article does not address the difference in side effects.)

"The depth of treatment in tissue for protons is related to a quantity known as the Bragg Peak. This is due to a buildup of dose in the final few millimeters of the proton range. The depth of the Bragg Peak is dependent on the energy of the beam; with increasing energy, the Bragg Peak is located deeper in tissue. ... As you can see, [See Figure that follows] *the entrance dose is relatively low, but as the beam penetrates deeper in tissue, there is a sharp rise in dose deposited. This is followed by a rapid stop in dose deposition. The beam stops at this point. Thus no tissue is treated beyond the Bragg Peak. This peak needs to be "spread out" to fit the width of the target to be clinically useful. Thus a special wheel, called a modulator is placed in the beam to spread out the Bragg Peak to the desired size.* [The Figure also] *shows a spread out Bragg Peak* [and also] *shows the relationship between an unmodulated Bragg Peak, modulated spread out Bragg Peak, and standard X-rays. Extensive studies have been performed to determine the biologic differences between protons and X-rays. A standard measure called the relative biologic effect (RBE) is used to compare the biologic effects of various radiation sources. A RBE of 1 is seen for standard X-rays. ... It turns out protons can be thought of exactly the same as X-rays in terms of its biologic effects because the calculated RBE is 1.1. Another measure of effect in biologic systems is the oxygen enhancement ratio (OER). Again, there is no difference in OER between protons and standard X-rays.*

The bottom line is that the only difference between protons and standard X-rays lies in the physical properties of the beam and not the biologic effects in tissue." [32] [End Article.]

The following diagram [33] shows a comparison of the relative dose of radiation received with a typical proton beam treatment of prostate cancer at greater than 200 MeV (million electron volts) versus "normal" X-ray or photon radiation at about 22 MeV. Keep in mind that it is the total energy deposited in the prostate gland that kills the reproductive ability of the cancer cells. Also, the X-rays and electrons are highest in some other area than the target prostate, and continue on to other non-target areas of the body. This is why the protons have few or minimal side effects compared to other forms of radiation used in cancer treatment.

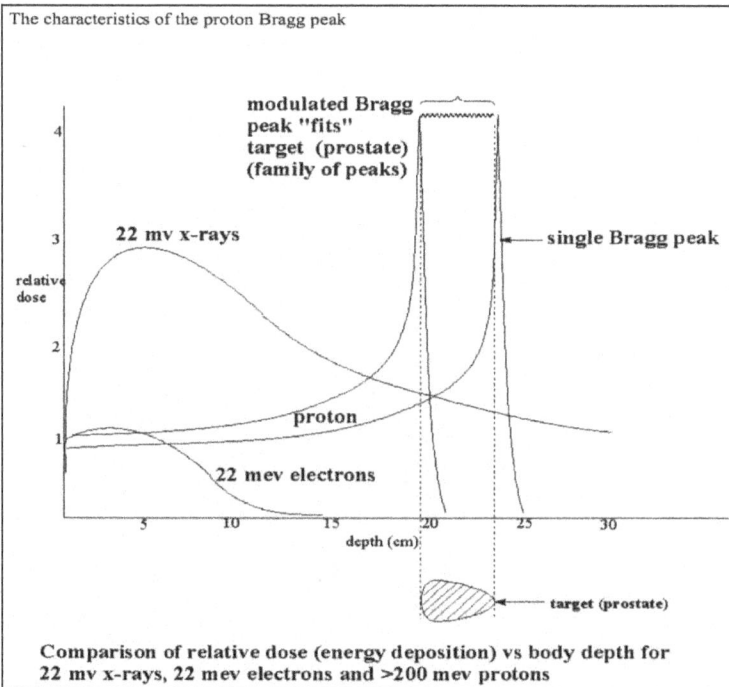

The characteristics of the proton Bragg peak

modulated Bragg peak "fits" target (prostate) (family of peaks)

22 mv x-rays

single Bragg peak

relative dose

proton

22 mev electrons

depth (cm)

target (prostate)

Comparison of relative dose (energy deposition) vs body depth for 22 mv x-rays, 22 mev electrons and >200 mev protons

It is this difference (what I call the "elegant physics of the proton") that makes the proton the superior radiation method for most cancer treatment when the cancer has not metastasized.

Yet another article found on the Internet is worth providing here for those without on-line capability:

"More than 200,000 men each year in the United States are diagnosed with prostate cancer. Until recently, the fear of treatment options rivaled the anxiety of the disease itself. In fact, the two primary choices of treatment, conventional radiation therapy, or surgery, concerned men so much that a third non-treatment option is often taken. It is called 'watchful waiting.' [Or Active Surveillance.]

But now a more advanced form of radiation therapy, without most of the side effects and limitations of conventional radiation therapy, is proving to be a very effective treatment for men suffering from prostate cancer. It also presents a viable alternative to watchful waiting.

For the past seven years [now seventeen]**, proton therapy, a superior type of radiation therapy that permits a more precise delivery of a higher dose of tumor destroying energy, has been successfully used at the Loma Linda Cancer Treatment center in Southern California.** *A recent study at Loma Linda University offers new information and hope in the battle against this devastating disease. The study concludes that when* higher doses *of conformal proton therapy are delivered at the target site,* **the results show a 'low incidence of side effects' compared to conventional radiation.**

The study further shows an extremely low rate of recurrences within the prostate treatment field. The result demonstrates that protons can be precisely delivered to the tumor, without causing most of the side effects of conventional treatments, and that the amount of energy delivered to destroy the tumors can be safely increased to enhance cure rates.

Proton therapy ... has a surgical precision that leaves vital organs and healthy tissue near the tumor unaffected, allowing for a speedier recovery for most of the patients." [34]

However, there still remains a great deal of controversy among various radiologists, oncologists, administrators, and other professionals regarding protons versus photons. The latest of these arguments center on the comparative costs of proton treatment and other radiation such as IMRT, and also the tremendous cost of new proton facilities. I believe that the cost of proton therapy will decrease as more facilities are built, and also that when smaller proton machines become available, these arguments will evaporate. There is no doubt that improvements in equipment and techniques for both methods will continue. Proponents of conventional photon or X-ray treatment argue that their treatment is as effective as proton therapy. One such doctor says: *"...the* [proton] *treatment has not been proved more effective than other cancer treatments such as intensity modulated radiation therapy and image guided radiation therapy, that are offered by competing hospitals."*

The above statement is one that I consider most debatable and subject to further clinical trials and expert studies with peer reviews. But—and this is most important—the above statement does *NOT* address the *potential for negative side effects* that could be experienced from the conventional radiation treatments, including IMRT. Nor does it address the associated costs of treating these side effects and the improved results (both efficacy of the treatment and lack of damage to surrounding tissue) obtained from proton beam therapy.

The above is only my opinion as a layman, but recognized experts in the field of radiation oncology agree. I provide two examples on the following pages.

The first of these is by Dr. Eugen B. Hug, former Director of Radiation Oncology at Dartmouth-Hitchcock Medical Center and a respected author of many radiation medicine scholarly articles, has the following, which he presented in 2004:

"Protons versus photons: a status assessment at the beginning of the 21st Century" (Dr. Eugen B. Hug.)

*"At present modern photon delivery techniques permit high dose isodose conformality similar to protons in many cases. However, **a proton advantage appears to still be present for target volumes of higher degrees of complexity and concavity.** Also, for selected histologies and sites, notably skull base tumors, **protons have established a "gold standard"** and photon data have yet to duplicate those proton results in clinical practice. Proton radiation therapy offers superior dose distribution by reduced low-dose integral irradiated volume."*

[This means total non-cancerous tissue that receives radiation is less.]

*"The avoidance of functional and cosmetic side effects in children by protons is at present acknowledged by the radiation oncology community and **is expected to gain similar recognition in adult patients**. This advantage of protons and disadvantage for photons constitutes an 'inherent physical gap' that will likely be long lasting. Although the priorities of proton irradiation advantages have shifted over the decades, **clinical advantages** [of proton therapy] **remain and are of sufficient significance and importance to justify further development and installation of proton radiation facilities worldwide."* [35]

[Bold emphasis added.]

Dr. Hug's assessment, given several years ago, was absolutely on target and correct. Two new proton facilities were brought on-line in 2006, and proton radiation cancer treatment centers are being built at an increasing rate both in the United States and worldwide.

Dr. Carl J. Rossi, whose work was referenced earlier, (and who was my attending physician at Loma Linda) recently provided a more up-to-date summary of proton radiation therapy status, particularly as used at Loma Linda University Medical Center's Department of Proton Radiation. In it he addresses the specific issue of protons versus IMRT. I have excerpted that portion because his explanation is exactly on point:

Protons

Protons VS IMRT (Dr. Carl J. Rossi.)

"One of the most recent technological advances in x-ray therapy has been the development of Intensity Modulated Radiation Therapy (IMRT). In IMRT, a computer-controlled linear accelerator is used to treat a tumor, like prostate cancer, with a multitude (the typical plan utilizes seven to fourteen beams) of individually shaped x-ray beams. In order to protect normal tissue, the intensity of the beams is adjusted dynamically by placing within the beam an absorbing material (typically made of Tungsten or some other high-density metal) so that when a particular beam path traverses large amounts of normal tissue the beam intensity is decreased. This type of treatment will produce a high-dose region which nicely surrounds the prostate but, (and this is an important point regarding the difference between proton therapy and IMRT treatment) **at the expense of increasing the amount of normal tissue which receives low to moderate radiation doses.** *The reason for this gets back to the aforementioned physical differences between protons and x-rays.* **IMRT is still fundamentally x-ray based therapy,** *and while one can change the intensity of the beam,* **what one cannot do (for it is a physical impossibility to do so) is to get an x-ray beam to stop at some point in space.***

[This is exactly what the "elegant physics of the proton" does with proton beam radiation therapy because of the Bragg Peak phenomenon. Continuing with the article:]

"Patients and colleagues routinely tell me that IMRT has made protons unnecessary because with IMRT we can achieve high dose distributions that in some cases are similar to those achievable with protons.

When I point out that the integral dose (total dose to normal tissue) is 3-5 times less with protons than when IMRT is used, they will often dismiss this low-dose radiation as being unimportant because it is not likely to cause any clinically identifiable problems. I consider this argument flawed because of the following points:

1. Virtually every advance in radiation treatment technology since the inception of radiation therapy has been directed at the goal of reducing any radiation dose to normal tissue to the greatest extent technologically possible. **IMRT reverses this trend because it exposes large amounts of normal tissue to low-dose radiation.**"

2. Our knowledge of radiation-induced organ injury is based primarily on analyzing the volume of that organ which receives a high radiation dose. Our understanding of the effect of treating a large volume of a normal organ to· a low dose is very limited, **and the long-term effects of such exposure are poorly understood.**

3. The only absolutely safe dose of radiation that we know of is zero. **Therefore, anything we can do technologically to limit as much normal tissue to zero dose will always be desirable and advantageous to the patient.** [The LLUMC Adventist philosophy again emerges!]

Conformal proton beam radiation represents the future of external beam radiation therapy. In terms [of] technological development it is presently in what I would consider to be the first generation of development. As accelerator technology matures, and active, scanned proton beams become a reality (this latter development is nearing its introduction into clinical oncology and is already available at one center in Europe) **the disparity between what can be achieved with protons and what one must accept with x-rays (including IMRT) will only become greater.**" [36]

[End Dr. Rossi's comments. Bold emphasis added.]

It is evident in any review of pertinent medical reports and publications that great differences of opinion remain regarding prostate cancer treatments and related efficacy and expected outcomes. It seems to me that this is because the overall outcomes related to "cure" rates seem to be reasonably similar. *Not so evident are the differences in quality of life issues in the comparison of potential side effects of the various modalities.*

But improvements in techniques, and the use of the various treatment methods, and combinations, continue.

Protons

The following is an abstract addressing the use of ADT (which is also AST–Androgen Suppression Therapy), in combination with radiotherapy that illustrates this point: *"The treatment of clinically localized prostate cancer is controversial. Options include radical prostatectomy, external-beam radiation therapy (EBRT), brachytherapy, cryotherapy, and watchful waiting. METHODS: The author reviews EBRT as treatment for clinically localized prostate cancer, with particular emphasis on the technological advances that have allowed dose escalation and fewer therapy-related side effects. RESULTS: Technological advances in the last two decades have significantly improved the delivery of EBRT to the prostate. This has resulted in an overall increase in the total dose that can be safely delivered to the prostate, which has led to modest improvements in biochemical outcome. An alternative approach of combining androgen suppression therapy and EBRT has also been successful in improving clinical outcomes. However, establishing the optimal therapy for prostate cancer remains controversial."* [37]

The radiation oncologists at LLUMC are well aware of such studies and improvements. Dr. Carl Rossi had me add an additional Lupron shot during my proton therapy (I rejected an additional one-month shot).

In my opinion, much of the discontinuity and differences that are prevalent regarding the various treatment modalities (especially the various forms of radiation treatment, including protons), are due to the large investments in the equipment for the various forms of treatment by hospitals and other facilities. Monetary investments must be recovered! There is also the fact that many specialists rely on the other modalities (brachytherapy, cryotherapy, surgery, etc.) for their means of livelihood. These specialties are what they have been trained to do, and medicine, like it or not, is a business.

As stated previously, the practice of medicine is a business, and business is primarily about making money. This is especially unfortunate in the case of the many segments of the medical profession, but nevertheless, true.

As Dr. Rossi so eloquently states above, proton technology is evolving. Only continuing improvements in technology, time, and perhaps the reduction of the cost of proton therapy facilities with a corresponding increase of the number of facilities and the specialists to operate them will mitigate the differences between proton and non-proton facilities. Of course, medical and scientific research continues, and further improvements in technology with perhaps better forms of cancer therapy (or even medicines) may appear and make even "The Beam" obsolete! One such possible development on the distant horizon (at least ten years) is the use of antiproton radiation for cancer treatment. Experimental work has already been done (2006) at the European laboratory for Particle Physics at Geneva, Switzerland (CERN), which shows that antiprotons may be up to four times as effective in the irradiation of cancer tumors as protons. [38] There is no doubt that when some of these new and expensive advancements occur, resistance to such new treatments will exist, (especially among administrators), because of the tremendous investments that have been and are now being made in proton facilities.

The fact that there are only five proton centers in the United States is due to the enormous costs of the physical facilities. Current costs for a proton facility, including the cyclotron or synchrotron and supporting equipment, approaches or exceeds $120 to $200 million in today's dollars. The new "Room Size" accelerator systems being developed by MIT and by UC-Davis (see "Status—" later) will hasten the availability of protons to all.

A current status of proton beam therapy facilities in the United States is provided at the end of this chapter.

Loma Linda Proton Radiation Facility Description

At Loma Linda University Medical Center, the massive synchrotron, gantries, and the beam transport system equipment reside underground, in the basement or "B-Level" Proton Radiation Center. At one end of this facility, the accelerator or synchrotron is located behind fifteen-foot thick concrete walls, along with the huge magnets and power supplies that control the proton beam. The synchrotron is about twenty-four feet in diameter, and is basically a ring of powerful magnets that control the protons within a high-vacuum steel tube. The energy capability of the machine is determined by design criteria, and at Loma Linda, the protons are accelerated to the desired energy or velocity, which can be from approximately 70 to 250 million electron volts,[39] and approach approximately two-thirds of the speed of light. The power supplies that energize the huge magnets (that weigh up to 7,000 pounds each) require a tremendous amount of cooling. They generate up to two megawatts of power; this requires additional supporting equipment.

The initial process starts when protons are stripped from the nucleus of hydrogen atoms in what is called an injector, and diverted to the synchrotron. After the synchrotron ring accelerates the protons to the desired energy, the beam is diverted by a special magnet system through evacuated tubing (the Beam Transport System), which has bending and focusing magnets that guide the beam around corners on the way to the treatment rooms, and maintains the protons in a highly focused beam (hence proton beam) within the vacuum tube.

This highly complex system is computer controlled and continuously monitors the intensity, size, and location of the beam at several points. Any variations from the prescribed parameters cause automatic adjustments, and automatic interlocks will shut the beam off in the unlikely event of failures.

There are four proton treatment rooms at Loma Linda, including one "Fixed Beam Room," which is the terminus of the horizontal beam line (HBL). This room actually has two beam lines, and is closest to the synchrotron. One of the two lines in the HBL is smaller and is used for certain eye tumors. The other is similar to the remaining three rotating head gantries, but is "fixed" horizontally. This HBL room is in regular use for prostate cancer patients along with the other treatment rooms, and the set-up is exactly the same for individual patients.

In the HBL, where I received all forty-four of my proton treatments, the patient alternates position each treatment day, first receiving the beam from one side, then from the other side the next day. Initial positioning is done by radiograph correlation to the positioning CT Scans done during the patient's preparatory phase. Correct alignment of the target prostate, using new radiographs each day, is done within (approximately) one millimeter.

The remaining three treatment rooms have huge rotating "Gantries" (approximately ninety tons each, three stories high—think mini-"Ferris wheels"—constructed of heavy structural steel) that allow the beam nozzles to rotate with them. This permits aiming the beam at a particular angle relative to the patient, who lies horizontal in his immobilization device or "Pod," at the center of rotation (the "iso-center") of the gantry wheel. This capability is useful for various other forms of cancer where full body access is needed. However, for prostate cancer the normal protocol is for the beam to enter the patient's body horizontally, attacking the prostate gland directly through the hips exactly the same as in the HBL. In the Gantry rooms, the patient's treatments alternate from the right and left sides of the body also, so the treatment is the same in each of the four rooms for the routine prostate patient.

There is an additional advantage of the rotating gantry beam nozzle. In the event that a patient had received a hip transplant or other metal implant that would interfere with the horizontal direct approach, the nozzle may be rotated to another angle, and still provide the radiation to the prostate or other cancerous target. Gantries 2 and 3 have automated digital X-ray set-up equipment—used for the initial positioning and alignment—which shortens the overall treatment time to a small degree. Gantry 1 has undergone modifications to permit robotic patient positioning prior to treatment, and the others will follow.

"If the system was built in 1990, is it up-to-date and comparable to the new facilities at other locations?"

This question was raised by one of the new patients during a Saturday guided tour of the proton system and facilities. His concern was because the installation was completed in the early 1990s, and almost new systems are in operation in Texas and Florida.

The answer, given by the man who is in charge of maintenance of all LLUMC proton systems, was that the present system is vastly improved over the original system, and when there are improvements that need to be incorporated, planned sequential shutdowns are done so that the work can be accomplished. This is in addition to the planned maintenance (the system is normally shut down on weekends). We witnessed such a repair during the tour. A large water pump, part of the cooling system, was being replaced. The overall design of the system is such that continuous improvements may be incorporated.

"The synchrotron and its proton-delivery system are designed to improve with time. They are to be modified as new technology becomes available." [40]

The last destination for the Beam Transport System is the "Experimental" Room, where various scientific experiments are carried out using the high-energy protons.

LLUMC is working with NASA as well as other medical institutions in basic research using this facility. The NASA research is targeting the effect of high-energy proton particles in humans during extended space travel. From a LLUMC Web site: [A previous] *"NASA administrator, Daniel S. Goldin, said the agreement represented "the ultimate in technology transfer,"* and further stated that the Loma Linda facility was the only place on Earth where NASA could do *everything* it needs to do to learn how to protect its astronauts from the dangers of positively charged particles in space. He went so far as to claim that Loma Linda's proton facility was NASA's "gateway to Mars;" [NASA is planning a manned trip to Mars].

Dr. Baldev Patyal, Chief Medical Physicist of LLUMC says: *"The main objectives of this [NASA] agreement are to provide access to accelerated proton beams and related research laboratories for NASA-sponsored investigators, to provide for contribution of NASA-sponsored investigators to the academic and educational programs of LLU, and to facilitate transfer of technical expertise between NASA and LLU in areas of radiation physics and radiation biology."*

"The scientific advances made at Loma Linda University Medical Center in the use of radiation therapy are of direct application to NASA's research into the biological action of energetic charged particles. Conversely, the results of the study of molecules, cells, and animals in space may lead to developments of importance to radiation medicine at LLUMC."

Note: I was unable to ascertain the status of this research project in April 2008.

Proton Beam Centers for Cancer Treatment
A Status Summary Update ~ April 2008
By Fuller C. Jones

The primary advantage of proton treatment for cancer is the precision of the proton beam and control of the dosage delivered to the tumor site. Unlike the standard radiation treatments that are available almost anywhere in the U.S. and in most other countries, the proton beam can be delivered directly to the cancer itself with significantly reduced or no damage to surrounding healthy cells, tissue or organs. This is of particular advantage to prostate cancer patients with contained disease that has not spread; but *many other cancers can also be treated*, including head and neck tumors, eye tumors, certain lung cancers, abdominal cancers, and breast cancers.

Because the proton beam treatment is not available except in a very few locations, it is not well known. It is usually not recommended except in the centers where it is available or nearby areas where publicity has made it known. For the prostate cancer victim and others, this is now changing. The recognition of the availability of the treatment—and acceptance as a viable alternative to surgery, brachytherapy (seed implant), cryosurgery, and standard radiation treatment—is becoming more widely known, particularly for prostate cancer patients where the tumor is still confined to the prostate capsule. More and more cancer patients are discovering the non-invasive proton beam, as results are made public, and patients that received the benefits of proton therapy spread the word. Reported results indicate at least a comparable record to other treatment methods, but without some of the side effects, such as incontinence and impotency. *"The patient feels nothing during treatment... [and] experiences a better quality of life during and after proton treatment."* [41]

Recent long-term reports of treatment history and results have generated a rapid proliferation of planned "Centers of Excellence" and primary medical institutions that are investing in the extremely expensive facilities to administer the proton beam therapy. In 2005, there were only three such primary proton beam medical facilities in the U.S. There are now (April 2008) five such centers with fully operational proton facilities that are currently treating cancer patients in a hospital environment: Loma Linda University Medical Center (LLUMC) at Loma Linda, California; Northeast Proton Therapy Center at Massachusetts General Hospital (MGH) in Boston; Midwest Proton Radiotherapy Institute (MPRI) at Indiana University, Bloomington; the M. D. Anderson Cancer Center in Houston, Texas; and the University of Florida Proton Therapy Institute (FPTI) at Jacksonville, Florida. One other, the facility at the University of California at Davis, treats eye cancer only. Worldwide, I have found references to twenty-eight Proton Centers currently treating patients, but some of these are in a laboratory setting rather than in a hospital-like environment. As of April 2008, about 50,000 proton treatments had been made worldwide.[42] (See the Endnote reference and link for the centers.) Of the U. S. hospital centers with dedicated proton treatment, the Loma Linda University Proton Center has been in operation the longest (since 1990), and for almost ten years stood alone. Loma Linda currently has the highest patient treatment capability, and can process between 125 and 175 patient treatments per day. It is significant that as of April 2008, approximately one-fourth of all cancer patients worldwide who have been treated with protons were treated at the LLUMC Slater Proton Center. In the case of prostate cancer treatments at LLUMC, normally forty-four or forty five treatments are required for a complete proton-only protocol, at 79.2 to 81.0 Gy[43] delivered at 1.8 Gy per day.

In some cases, where there is a chance that cancer has spread within the prostate bed, photon (X-ray) treatment is also required and the proton protocol may be varied.

As of April 2008, the LLUMC Proton Center has treated well over 12,000 cancer patients with many types of cancer disease. More than half of these (actually about 65 percent) were victims of prostate cancer. The Texas and Florida centers have only been in operation for a short time (since mid-2006), but are now fully operational. There are set-up variations at the different locations, depending on the facilities and doctors. However, the daily use of protons in the hospital environment has been proven (at LLUMC and the other active proton centers), and proton treatment protocols are well established.

Highlighting the growing recognition, progress, and degree of potential for proton beam treatment, there are several new centers either under construction or in the advanced planning stage within the U. S., most requiring an investment of $120 million to $200 million.

The University of Pennsylvania is building a large facility near Philadelphia, which is being partly funded by The Dept. of Defense in partnership with Walter Reed Army Hospital.[44;45] Construction of this facility is well underway. The cyclotron, built by IBA of Belgium, arrived in Philadelphia January 29, 2008.

Hampton University in Hampton, Virginia, is planning a $183 million facility (groundbreaking has taken place). The 98,000 square foot facility is scheduled to open in 2010, and will treat approximately 125 patients daily (over 2,000 patients per year). It will feature *four* gantry rooms and one fixed beam room. Most other new centers have only two or three gantry rooms.

ProCure Inc., the developer of the Bloomington, Indiana facility, is constructing a private (for profit) Proton Center in Oklahoma City[46; 47] that is planned for 2009. Another Procure project (full size, $159 million) is now planned in partnership with Beaumont Hospitals of Michigan, but regulatory permissions are needed.

Oklahoma City may have a second proton center. A follow-on announcement stated that Oklahoma University Cancer Institute is to build a proton center on the Health Center campus, as part of the new university cancer center.

In October 2006 Northern Illinois University announced plans to build a world-class cancer treatment and research center in that will provide state-of-the-art proton therapy.[48] The facility will be known as the Northern Illinois Proton Treatment and Research Center. Central DuPage Hospital of Winfield, Illinois, a suburb of Chicago, is also pursuing development of a proton center.

The Seattle Cancer Care Alliance is planning a $120 million center in Seattle, Washington. A Letter of Intent was signed in February 2008; therefore this facility will probably not be on-line until late 2010 or 2011.

Barnes-Jewish Hospital in St. Louis, Missouri; Broward General at Ft. Lauderdale, and Orlando Regional at Orlando, Florida, are planning smaller units ($20 million; see reference to MIT proton development below) to be brought on-line in 2009 and later. There are about fourteen others in the proposed, pre-planning, or design stage in the U. S. and worldwide.[49] Experts foresee up to 100 U.S. proton centers within the next few decades.[50]

It is quite evident that proton beam radiation therapy for cancer treatment is a treatment modality whose time has finally arrived.

There are on-going improvements in the present technology. LLUMC is now installing robotic positioning systems. Also at LLUMC and other locations "Image Guided" and "Active Scanning" proton delivery devices are being planned that will enable even more accuracy in proton delivery to target tumors. Some of these are already in use at some locations in Europe.

Protons

The future promises even more exciting developments. The great hindrances to universal use of protons in cancer treatment are the size and cost of the cyclotron or synchrotron equipment and supporting facilities necessary. The Massachusetts Institute of Technology (MIT), in collaboration with a private development company, is working on a comparatively small (room size) accelerator to deliver the proton therapy to patients. If this development is successful, an even more rapid expansion of proton beam therapy should almost immediately occur. According to the MIT News Office:

"MIT proton treatment could replace x-ray use in radiation therapy. Scientists at MIT, collaborating with an industrial team, are creating a proton-shooting system that could revolutionize radiation therapy for cancer. The goal is to get the system installed at major hospitals to supplement, or even replace, the conventional radiation therapy now based on x-rays. The fundamental idea is to harness the cell-killing power of protons.... Worldwide, the use of radiation treatment now depends mostly [approximately 90 percent] on beams of x-rays, which do kill cancer cells but can also harm many normal cells that are in the way. What the researchers envision -- and what they're now creating -- is a room-size atomic accelerator costing far less than the existing proton-beam accelerators that shoot subatomic particles into tumors, while minimizing damage to surrounding normal tissues. They expect to have their first hospital system up and running in late 2007." [51]

Note, this date was obviously optimistic; it is apparent that this machine will not be available until 2008 *or later*, with first patients treated about 2010. A search found that the MIT technology was licensed to Still River Systems of Littleton, Massachusetts. The trade name given to the MIT-Still River systems device is "Clinatron-250™."

As has happened many times in the history of modern technology development, there are others concurrently working on the idea of a smaller, less expensive proton accelerator. The University of California Davis Cancer Center is actively engaged with a similar project. In the Fall/Winter 2006 "Synthesis" (Volume 9, No. 2), there is the following:

"UC Davis Cancer Center and Lawrence Livermore National Laboratory join forces to make proton-beam therapy available to every major cancer center." The story goes on to describe the coordinated efforts of UC Davis and the Lawrence Livermore Laboratory to develop the machine:

"Size and cost have been the obstacles. A 90,000-square-foot building — bigger than many hospitals — is needed to house a state-of-the-art proton-beam accelerator. And the machines carry price tags of up to $150 million. But these barriers may be about to topple. Researchers from Lawrence Livermore National Laboratory and UC Davis Cancer Center are working on a subscale prototype of a "miniaturized" proton-beam accelerator. Led by George Caporaso of Livermore's Physics and Advanced Technologies Directorate, the research team aims to deliver a final machine that will be small enough to fit in a typical radiation oncology suite, powerful enough to treat cancer anywhere in the body and priced at about $10 to $15 million. The lab is currently seeking commercial partners to help construct a full-scale model." [52]

In the new technology transfer pact, Lawrence Livermore National Laboratory has licensed the technology to *TomoTherapy, Incorporated* of Madison, Wisconsin, through an agreement with the Regents of the University of California. Note: I suspect that the final cost of such machines may be closer to $20-$30 million. This is of course much less than the cost of the current large-scale proton facilities, and is more within the range of possibilities for most large metropolitan hospitals. I also think that some of these new systems will incorporate the "scanning" method of beam delivery. Proton technology is advancing so rapidly that it is difficult to keep track of new developments.

Further into the future are more exciting developments, such as using carbon ions or antiprotons, in the pursuit of new tools to fight cancer.

From a Web site called ACT, Advanced Cancer Therapy, "Committed to Increasing Knowledge of Advanced Cancer Therapies Using Particle Beams to Terminate Cancer Cells:"

"Which particles are used in advanced particle beam cancer therapy? Up to now the particle of choice was the proton, the nucleus of a hydrogen atom. 40+ proton beam therapy centers exist and have to date treated more than 50,000 patients worldwide.

In recent years research in Germany and in Japan has shown that carbon ions can have a much higher biological impact on cancer cells than protons and can therefore successfully treat tumors normally deemed "radio-resistant". Carbon ion treatments have yielded significantly improved treatment results in many types of tumors. But up to now only 3000 patients have received Carbon ion treatments.

Antiprotons, known to most of us only from science fiction, have already been shown to offer yet another increase of effective dose in the target area, have the potential to further decreasing the impact on healthy tissue in front of the tumor, and additionally would allow watching in real time where exactly inside the body the treatment is administered." [53]

One such project is in the not-so-distant future here in the United States. In July 2007, Touro University announced plans to build a center for particle therapy cancer treatment in California that will offer both proton and carbon ion therapy. Upon completion, it would be the first such center in the United States, and would be part of the school's future $1.2 billion health science research campus on Mare Island in the San Francisco Bay area.

CERN, the European Organization for Nuclear Research, has been a pioneer in high-energy particle research, and continues in this effort along with members that include most of the world's developed countries.

CERN is actively investigating the use of antiprotons to fight cancer. ACE (Antiproton Cell Experiment) is the first investigation of the biological effects of antiprotons. The research has shown that antiprotons are up to four times as effective as protons in eradicating cancer cells. According to Michael Holzscheiter, who is the spokesperson of the ACE experiment:

"To achieve the same level of damage to cells at the target area one needs four times fewer antiprotons than protons. This significantly reduces the damage to the cells along the entrance channel of the beam for antiprotons compared to protons. Due to the antiproton's unsurpassed ability to preserve healthy tissue while causing damage to a specific area, this type of beam could be highly valuable in treating cases of recurring cancer, where this property is vital."

~~~

This "Status" summary would not be complete without acknowledging the recent (within the past year) surge in the number of cancer patients applying to the various proton centers for treatment, especially prostate cancer patients. In less than a year, LLUMC in California has developed a four to five month backlog of applicants. About a three month backlog is now (April 2008) confirmed at the M. D. Anderson facility in Houston, Texas. Massachusetts General Hospital in Boston has essentially severely restricted their acceptance of prostate cancer patients. So far, FPTI in Florida and MPRI in Indiana are accepting prostate cancer patients for consults, and providing treatment within reasonable time spans (a month or so).

No doubt as this trend continues further restrictions will occur. Only the activation of several of the large new proton facilities (Philadelphia, Chicago, Oklahoma City, and Hampton, Virginia) will help to mitigate this situation. Protons are in demand!

The Legacy of Loma Linda

IF I TRIED TO INCLUDE everything that I have discovered about Loma Linda University Medical Center, this chapter would be a book in itself. So I must summarize as much as possible.

In 1975, Richard A. Schaefer first published *"Legacy ~ Daring to Care, The Heritage of Loma Linda University Medical Center."* The 2005 revised edition contains many additions added to reflect the "pushing the envelope" medical advances pioneered by this remarkable institution. In reading and studying Richard's book, I found many examples of an institutional and medical staff that was not content to "rest on its laurels," but continually tried to forge ahead where the need existed. You may be able to obtain a free copy of the book from LLUMC Media Relations.

In any great endeavor, leaders of extraordinary vision and capability are required. LLUMC has been blessed with a continuation of such leaders from the beginning until the present day! All seem to have had (and still have) great vision and remarkable faith. Loma Linda University Medical Center is, first of all, a religious institution that puts faith in God as an over-riding priority. I believe that this must play a major role in the overall success of this great medical center.

In Richard Schaefer's book the astounding history emerges. Loma Linda University was formed from the visions of a woman whose name was Ellen White. I found this to be quite literally true. Richard provides the details complete with stories and documentation. The result is a remarkable history of an institution that was founded on religion and faith. What a labor of love is his work!

Chapter Ten

In his words: *"The story of Loma Linda is really a saga—a story of heroes and heroines...a story of faith...of mission...of daring...of continued daring. It is the story of the deep desire to serve God and humanity, a story that is rich in compassion and vision..."* [54]

Much of the following material is drawn from Mr. Schaefer's work, which he has kindly permitted me to reference in this chapter in a general way. Several specific citations are shown. Following the third reference, citations for *"Legacy"* will be provided in the text.

Loma Linda University Medical Center had its genesis in a strange accident. In the 1830s, when Ellen Harmon was a young schoolgirl, she was struck in the face by a rock. This unfortunate incident severely disfigured Ellen and traumatized her to the point where she was unable to continue her formal schooling. She received less than three years of education. In spite of this disadvantage (or perhaps because of it), Ellen became intensely religious and turned to self-study of the Bible. Her family belonged to a religious group founded by William Miller, whose belief was that the second coming of Christ was at hand. The Millerites, as they were called, actually issued predictions for the dates when this was to occur. One of these dates was October 22, 1844. When the foretold event did not occur, this date became known as "The Great Disappointment" among the Millerites. Soon after this, Ellen Harmon, a frail, shy, teenager stood up in front of a group of Millerites and announced that she had received a vision from God!

Between 1844 and 1846, Ellen Harmon continued to recount her visions and demonstrate her leadership abilities. She matured, and married James White, another church leader, who had heeded Ellen's counsel to avoid "date setting" because it was not God's will.

A parallel story also begins in the 1830s, when John P. Kellogg left New York to pioneer a settlement in what was to become Flint, Michigan. He bought 320 acres and proceeded to build a home for his family. After several years and moves, John Preston Kellogg finally moved to the area of Battle Creek, Michigan. During this time there were many "medical" mistakes made by what the practitioners thought was right in those times, that adversely affected the Kellogg family. One of these caused the death of his first wife, and another the death of a beloved daughter. Because of this, Mr. Kellogg and his family grew understandingly bitter. The reason that this is significant will be shown later.

The history of medicine as it was practiced in America during the 1700s, throughout the 1800s, and even to the early 1900s is remarkable in the terrible methods used to "cure" and "help" sick people. How this was changed and the role that the Adventist religion played in the changes is an important part of our history.

"Bleeding" and "Drugging" were primary treatments. Physicians adhering to the "Bleeding" school of thought believed that too much blood caused inflammation and fever. Therefore blood was drawn from the ailing patients to effect the reduction of fever and to "cure" whatever disease was causing the high temperature!

It is not commonly known, but bleeding hastened the death of George Washington. He lay very sick and almost unable to breathe when: "The bleedings inflicted by Washington's doctors hastened his end. Some 80 ounces of blood were removed in 12 hours (this is 0.63 gallons, or about 35% of all the blood in his body)." [55]

This practice of bleeding to "cure" patients was widespread. It is impossible to even imagine the number of deaths that occurred that were directly attributable to the loss of vital blood because of this horrible "treatment."

The (so called) doctors who believed in "Drugging" thought that as the patient's body fought to overcome the drugs, it would also overcome whatever disease he or she was afflicted with. Such things as opium, heroin, arsenic, prussic acid, and even strychnine, to name a few, were commonly used! Sometimes both bleeding and drugging were used on the patient. The "cure" rate of these early practitioners was extremely low! Richard Schaefer has a succinct comment: *"Rigor Mortis was a frequent complication in the practice of medicine."* [56] This period of medical history was fraught with these terrible practices, and quackery abounded.

In the 1800s unscrupulous traveling salesmen sold "Snake Oil," "Genuine Indian Remedies," and countless other such "cures" in every town and hamlet. They were also sold in the early "General" stores, and even in the venerable Sears and Roebuck Catalog! Many of these "remedies" contained alcohol or opium, so it should certainly not be a surprise that they were widely used.

So called "medical doctors" practiced "medicine" with no formal training and even without licenses. Anyone with an interest in medicine and a source for drugs could call himself a "Doctor." There was NO attempt to provide regulation during the 1800s and even in the early 1900s.

"Until early in [the 20th] *century any drug, no matter how worthless or outrageously adulterated, could be placed on the market. ..."* [57] "Home Health Guides," were commonplace and widely used, even though most of the recommendations in them were of no use and were even dangerous to use. One such recommendation was to use tobacco smoke for lung problems (by smoking a cigar)!

There was also little or no attempt to improve health practices. The fact that cleanliness was necessary to provide the germ-free environment necessary for healthy living was totally unknown. Sanitation in the daily lives and surroundings of most people simply did not exist. Sickness, contagion, and mortality rates from disease were very high. To put this in some perspective, in the mid to late 1800s, the infant mortality rate was very high; almost one in six babies did not live to the age of one year. Today this ratio is more like seven in 1,000. And in 1850 the average American lived only until about 39 years of age; now this is around 77, and this age continues to increase with advancing medical knowledge.

All of the foregoing is background information, and is provided to emphasize the importance of the influence of the Adventist doctrines and practices on the improvement of health and the practice of medicine in the United States.

Health reform movements did not really start in the United States until the mid 1860s, when a group of Christians formed a new religious denomination, calling themselves "Seventh-day Adventists." The founders of this institution were following the counsel of a young woman whose name was Ellen G. White. This remarkable woman had demonstrated extensive knowledge and foresight in spite of no medical training. In fact, she had almost no formal educational training at all! The reason that the church elders followed her counsel was because of her demonstrated leadership abilities and her "visions" that many believed were messages from God.

In 1860, the church had heeded Ellen White's advice and counsel, and established the organized structure that later became the present Seventh-day Adventist Church. Ellen White, the wife of James White, was the same Ellen Harmon, the young girl whose education was cut short and yet whose visions and prophecy guided the organization and policies of the Seventh-day Adventist Church for the next forty-odd years! In spite of her lack of formal education, this remarkable woman wrote many books that became guidelines for modern health practices.

Chapter Ten

This Adventist group advocated Christian living and this included cleanliness and health care reform as well as Godliness. They attempted to make these practices known throughout the members of their congregations, and also within the communities in which they lived.

The new church opened the first Adventist medical institution, "The Western Reform Health Institute" near Battle Creek Michigan in 1866. Richard Schaefer says it best: "The founders of the church wanted an institute that, unlike any then existing, would combine all of the best reforms—an institute that would use the best simple, natural remedies and the best surgical and medical procedures of the day, rather than the era's poisonous drugs; *an institute in which the spiritual well-being of the patient would be the object of as much concern as the physical well-being....*" (Schaefer, *"Legacy...."* p. 79. Italicized emphasis added.)

This first Adventist medical facility was started by a $500 donation from John Preston Kellogg and another donation of the same amount from Ellen White! John's donation was motivated by his personal knowledge of the terribly poor medical help that was currently available. It had cost him his first wife and a beloved daughter. Ellen made her donation because of her personal vision, convictions, and faith. She saw with great clarity the benefits of the new hospital and the future successes that were to follow.

The success of this new medical institution during the following years established a blueprint for medicine and health reform in the rapidly growing new country of the United States of America. Patients, patrons, and guests at this hospital read like a "Who's Who" of the early 1900s, with names like Henry Ford, William Howard Taft, Harvey Firestone, John D. Rockefeller, Bernard Shaw, William Mayo, J. C. Penney, and Amelia Earhart, as famous guests or patients.

The reputation of the new Adventist hospital expanded internationally, especially after Dr. John Harvey Kellogg became associated with the hospital. Dr. Kellogg was also an inventor and industrialist as well as a highly respected physician and surgeon. His reputation as a surgeon spread throughout the United States because of the successes of his operations. And among his many inventions were peanut butter, flaked cereals, and "exercise machines." This Dr. Kellogg, yet another example of a strong visionary leader of the Adventist Church, was the son of John Preston Kellogg, whose "seed" money started the funding of the institution.

The Adventist doctrine continued to advance the cause of health reform and better medical practices throughout the 1800s. In the late 1800s, leaders of the church saw the need for more physicians trained in the Adventist philosophy as well as the standard medical methods. This led to the formation of the American Medical Missionary College (AMMC) in 1895. Coincident with this event the medical leaders announced that the new college was NOT a sectarian school. In spite of counsels from Ellen White and other Adventist Church leaders, the medical leaders proceeded to declare the new college "undenominational."

Despite the controversy, the medical missionary work of the Seventh-day Adventist Church continued to expand to other countries all over the world.

There were setbacks through the years, such as the fire that completely destroyed the Battle Creek Sanitarium in 1902. This precipitated a series of events that threatened to destroy the Adventist Battle Creek efforts, and even threatened the Church itself.

The medical leaders disagreed with the Church, and wanted to rebuild the hospital in a larger and grander manner. They also wanted to make the new hospital totally non-denominational. These separatists, primarily the medical leaders, eventually separated themselves from the Adventist Church, and built the replacement in a grandiose and very expensive manner. This was in spite of warnings from Ellen White that the hospital should be rebuilt in a smaller, simpler way, with a "Christian atmosphere," and should consider the needs of the patients first. The separatist leaders did not do this, and instead built on a grand scale, with many extremely luxurious and costly embellishments. Several years later the (non-Adventist) hospital failed financially, went bankrupt and was in receivership, along with sister institutions in Chicago and Miami.

The American Missionary Medical College continued to provide well educated, proficient, and Christian professionals until 1910, when it merged with Illinois State University. (Schaefer, *"Legacy...."* p. 134.)

As the old AMMC merged with the state college and ceased to exist, providentially, a new era began. By the very early 1900s, the Adventist Church had established 27 sanitariums and "treatment rooms" in the United States and in foreign countries (as part of their missionary movements). In 1904 and 1905, with Ellen White's guidance and insistence, new sanitariums were established in Southern California at Glendale, Paradise Valley, and Loma Linda.

In 1910, a totally denominational Adventist medical teaching institution, the College of Medical Evangelists, opened in Loma Linda, California. The events leading up to this are carefully described and documented in Richard Schaefer's book. If it had not been for Ellen White's vision, leadership, persistence, and never failing faith, the many faceted Loma Linda University Medical Center that we know today would never have come into existence.

The remarkable manner in which this happened began in 1900, when a group of business men and doctors started the "Loma Linda Association" for the purpose of creating a new medical facility utilizing a failed luxury hotel of 64 rooms and the surrounding land.

They did so, but by 1905, this well equipped medical facility had failed and was for sale. Ellen White's vision had foreseen the availability of the property and even the description of the buildings and surrounding territory, even though she had not actually seen it with her own eyes. In spite of a lack of sufficient funds (even though the property was offered at the fantastically low price of $40,000.00), Mrs. White prevailed upon the Church leaders to attempt to acquire it. Funds were providentially made available, and in August of 1905, The Loma Linda Sanitarium, a Seventh-day Adventist medical institution, was officially formed. Mrs. Ellen White continued her determined and forceful leadership, insisting that the institution must be an educational facility as well as a healing place. As the early years passed, her visions continued to bear fruit; in 1910, she said: "The medical school at Loma Linda is to be of the highest order ..." (Schaefer, *"Legacy...."* p. 152.)

The growth of the institution continued in a providential way, despite financial, legal, and the initial obstacles thrown up by the American Medical Association. Finally, in 1922, the school was awarded an "A" grade by the AMA. The University continued to grow and expand.

Eventually, in the early 1950s, a School of Dentistry was added, and since then several important advances in practice of dental medicine were developed and accepted.

In 1967, the medical school began the work that would eventually "put Loma Linda on the map," with a kidney transplant. This led to the establishment of a team of surgeons who were performing the first newborn heart transplants in the mid-1980s, and suddenly Loma Linda was recognized as advancing medical knowledge worldwide. The skills necessary have been honed to an amazing level. A baby receiving a heart transplant at Loma Linda now has about an eighty-seven percent chance of being alive after three years. This is better than most conventional heart surgeries for complex cases.

The development of a program to locate and transport infant hearts for waiting transplant recipients was a natural follow-on to the advancing expertise of the transplant team. Loma Linda transplant surgeons can now travel to any location in North America and the Hawaiian Islands to procure hearts, and return them to Loma Linda in time for the completion of the transplant!

The transplant work led to the establishment of the Transplantation Institute. Then came the Cancer Institute, the Rehabilitation Institute, the Orthopaedic Institute, and also the International Heart Institute, which has for many years been at the forefront of leadership in research and treatment of heart disease. Research and advancement of medical knowledge became objectives in addition to the over-riding "Make Man Whole" philosophy.

Yet another instance of this philosophy to "Make Man Whole," and "Pushing The Envelope" in medical science and treatment is a surgical procedure called Pallidotomy, which has been called a "medical miracle for Parkinson's disease". This neurosurgery is in fact brain surgery, and uses the most advanced equipment and techniques available. The device that really makes the procedure possible is the Magnetic Resonance Imaging Scanner, which allows the surgeon to see inside the patient's brain and locate the hyperactive brain cells that can cause the tremors and other debilitating effects of Parkinson's.

Through a small hole drilled in the patient's skull, a probe is inserted with extreme precision, using the MRI generated "navigational aid," into the globus pallidus of the brain. When the correct location is reached and verified, the tip of the probe is heated, causing a small lesion that blocks the abnormal brain signals that cause the effects of Parkinson's. The most astounding part of this procedure is that it is sometimes done on an out-patient basis, with the patient under local anesthetic, and awake during the complete procedure, which can take up to three hours! *The patient is able to see the improvement immediately—as it happens."* (Schaefer, *"Legacy..."* p.46.)

Robert Iacono, MD, associate professor of surgery, anesthesiology and neuroradiology, School of Medicine at LLUMC, has performed more than 1,700 of these procedures since 1991. While the procedure is not always a "cure", and is not recommended for all Parkinson's sufferers, it can alleviate most of the symptoms for those who qualify, usually those who do not respond well to medication.

And before this, Dr. James M. Slater, head of the Loma Linda Department of Radiation Sciences and Medicine, also had a vision. He foresaw a pressing need for the use of heavy charged particles (protons) in the battle of that dreaded disease, cancer. Here is yet another example of a daring and visionary leader who tirelessly worked for a dream. In 1971, Dr. Slater started a team to investigate the use of protons for the treatment of cancer. As a direct result, the world's first proton accelerator designed specifically for cancer treatment in a hospital environment became operational at LLUMC in 1990. It is difficult to imagine the struggles and problems that Dr. Slater dealt with in convincing the administrators to go ahead with this vast project. To appreciate the magnitude of the design and engineering tasks associated with the construction of the LLUMC Proton Center, one must view the recorded history of the construction, which is available free from the LLUMC Media Relations. The underground facility is indeed awesome, with the synchrotron and supporting equipment, and the massive three-story high 100-ton gantries. Having seen the actual hardware, and having been a prostate cancer patient on the receiving end of that proton machine, I appreciate it even more.

Proton beam therapy is advanced radiation medicine based on nuclear physics, and the LLUMC facility has treated thousands of cancer patients from all over the world. LLUMC was the first in the United States to use protons to treat patients in a hospital. There are relatively few proton treatment facilities anywhere in the world. It is very remarkable that, including all patients anywhere who have chosen protons, LLUMC has provided the treatment for over one-fourth of them.

What is also impressive, and attesting to the safety of the Proton Center that lies below it, is that the Loma Linda University Children's Hospital stands six stories directly above the concrete and steel shielded Proton Treatment Center facility. This close proximity is a sort of symbiotic relationship, because proton beam therapy is uniquely qualified to treat children with cancer.

"When doctors began using proton therapy in 1990 for the routine treatment of cancer patients at Loma Linda University Medical Center in Southern California, treatment strategies focused primarily on adults. Since then, new proton treatment programs have emerged that focus successfully on pediatric oncology as well. So much so that Loma Linda University is now the pre-eminent institution for treating children's cancer with proton therapy." [58]

"Proton radiation is so precise that doctors can target the radiation directly at the diseased site with almost no

damage to the surrounding healthy tissue. In this way, children with certain cancerous tumors can be treated with almost none of the side effects that usually come with traditional radiation therapy." [59]

The proton advantage is because children are much more sensitive to side effects of radiation than adults. This is because all of the young body's cells are rapidly growing, and any disruption of the normal progression of this growth, such as may happen with normal X-ray treatment, will cause later developmental problems. This is particularly true with tumors of the brain, and is why protons are the treatment of choice in children's brain cancers, indeed, for *any* cancers of children. As the use of protons for cancer therapy becomes more widely known and practiced, this proton advantage can only increase.

While at Loma Linda for my proton treatments I saw many children, of all ages, being admitted to the Proton Radiation Center, and others there for their treatments. It was heartbreaking to see some, who were obviously advanced cancer patients, but heartwarming to observe the efforts of the staff to keep them cheerful and upbeat, and to see the happy responses of the children. This was yet another instance of the demonstrated caring philosophy of the entire hospital.

As already discussed, the Children's Hospital is also recognized as the world leader in infant heart transplantation, and the neonatal (care of newborn children) unit is one of the largest and most advanced in the world. Providing the very best possible medical care for children is obviously of the highest priority at Loma Linda University Medical Center.

"Loma Linda University Children's Hospital is a comprehensive, state-of-the-art medical facility designed especially for children's unique health-care needs." [60]

In fact, the Loma Linda University Medical Center is now practicing the most advanced medicine in every clinical specialty, and has an international focus. As a teaching facility and a hospital, it is attracting teachers, students, physicians, and patients from every corner of the globe.

Loma Linda University Medical Center, which began as a small denominational sanitarium and school championed by one woman with a vision, is now an amazing medical and teaching institution that stands unique, and continues to advance the frontiers of science and medicine while maintaining the strong Adventist doctrines of healthful living and spirituality.

The Brotherhood

THE BROTHERHOOD OF THE BALLOON is an organization made up of prostate cancer patients who have chosen proton beam therapy as their method of treatment. Robert J. Marckini founded this organization in the year 2000 after he completed his proton treatment at LLUMC. In going through the daily routine of these treatments, Bob found, as most patients do, that comradeship and fellowship just "happens." Friendships are formed with other patients that are often of a lasting nature. It is natural to want to "keep in touch" with your comrades, if for no other reason than to see how successful the protons have been in fighting the cancer demon. In addition, it is natural to want to keep track of newfound friends, no matter where they are.

Bob and a few of his fellow patients decided to form a loosely bound "club" to facilitate keeping in touch, and called this group "The Brotherhood of the Balloon" for obvious reasons. This initial group was formed with only Bob and six of his new friends. After his treatment was completed, Dr. Lynn Martell and Gerry Troy mentioned the new group at the regular Wednesday night meetings. (Mr. Troy is now with the new University of Florida Proton Institute at Jacksonville.)

Mr. Marckini found that after a year or so, this "club" took on a life of its own. He describes this in his book, and I recommend that you read it. Here I will just say that the "BOB," or Brotherhood of the Balloon, now (April 2008) has over 3,000 members from all over the world! The organization has become practically a full time occupation for both Bob Marckini and his wife.

Membership in this organization is not automatic, however. Some men who have had the proton treatments assume that they are "automatically" members of the BOB. This is not correct. One must join, either by filling out a paper form or by doing so on-line at the ProtonBob.com Web site.

In addition to Bob Marckini's book, *"You can Beat Prostate Cancer, And You Don't Need Surgery To Do It,"* this Web site is a MUST READ for anyone diagnosed with prostate cancer or for anyone who wants to learn more about the disease or treatments. There is a portion of the site that is "open" to anyone, providing links to other sites that are extremely informative, and access to testimonials that are compelling to say the least. This is the part of the site that took up more than five hours of my late-night and early morning time when I first found "The Beam," and contributed greatly to my decision to go to Loma Linda University Medical Center for proton beam therapy treatment of my prostate cancer.

However, after joining, a password is issued that allows access to the "Members Only" portion of the site, and *much more* detailed information is available for research and study. The data on this site is extremely useful in helping a newly diagnosed patient in his decision process, in addition to post-treatment care for those who have already undergone treatment.

From the Web site: *"The BOB maintains a private database, which includes members' vital statistics, along with pretreatment cancer stage, follow-up PSA records and other information. The purpose of the BOB is to provide an aftercare communications forum, where members share information on any subject related to their treatment, the healing process, and preventing a recurrence. Members also promote prostate cancer prevention at public forums and willingly share their experience of Proton Treatment with others who have been diagnosed with prostate cancer, as a way of helping them decide among prostate cancer treatment options."* [61]

Another important and useful advantage to joining the BOB is receipt of the monthly newsletter, which Bob Marckini has named "BOB Tales." All members will receive this important communication, either by Email or (with the help of volunteers) by "snail mail" when computer access is not available. Bob describes the newsletter in his book, and here is a brief quotation:

"Every month we report on the status of new member additions; new member feedback; the Featured Member of the Month; prostate cancer prevention, detection, and treatment news; BOB Member Reunions; news from the major proton treatment centers; how we are doing on our mission to support each other, promote proton therapy, and give something back; health and nutrition suggestions and tips; and some humor to brighten their day." [62] The January 2008 issue has: *On December 12, 2007, the center became the* ***"James M. Slater, MD, Proton Treatment and Research Center."***

The Brotherhood of the Balloon has grown in leaps and bounds during the last couple of years, and continues to grow as more facts are made available that attest to the success of the modality, and that lead to "The Beam" at LLUMC or at other centers for treatment.

I have personally experienced the remarkable bonding that occurs between the members of this organization. It begins with the common realization that we are all human and subject to this dread disease, and we have all reached the same conclusion that the correct treatment for us is the Proton "Beam of Hope." When you consider that there are thousands of men with prostate cancer who have never heard of protons, we are indeed unique.

Chapter Eleven

The bonding grew as we met daily at various times for our "daily dose" of protons, and was strengthened during the weekly "Support and Education" meetings sponsored by LLUMC and further strengthened by the leadership of these meetings. During my stay at Loma Linda this leader was Dr. Lynn Martell, yet another visionary leader and strong Christian Adventist.

The humor, fellowship, and overall attitude of the patients exhibited during these meetings is impossible to describe; it must be experienced to be believed and understood!

Can you imagine a group of sixty to one hundred people, both men and women, from all over the world, whose entire lives had been disrupted and threatened by a dreaded and deadly disease, meeting weekly with laughter and fellowship, and all demonstrating the complete belief that they were absolutely in the best place in the world for their treatment, and were receiving it? I could not, until I experienced it!

As the number of Proton Centers increases, the Brotherhood of the Balloon will also grow. Even though the "Home" facility, Loma Linda University Medical Center, will become one of many locations where the Beam of Hope is used in the battle against cancer of all types, this remarkable organization will continue to be the "Men's Club" advancing the use of proton beam therapy for fighting prostate cancer.

It is my opinion that eventually the so-called "gold-standard" for treatment of prostate cancer will no longer be the radical prostatectomy that is now most often recommended by urologists (surgeons) all over the world. Instead, proton beam radiation therapy will emerge as the "best" treatment in most cases, regardless of the age of the patient. In my opinion, proton beam therapy will default as the new "Platinum" standard, and become recognized as such all over the world.

The Brotherhood

This may take a few years longer because of the fact that there are so few centers with proton treatment capability. But this is beginning to change, and the end result is almost inevitable. This may be a lengthy process, because it will be an educational process as well. The proponents of the other modalities will continue to express doubts about the proton beam, and many newly diagnosed patients will no doubt listen to them and not be convinced of the efficacy of the proton treatment. Many other patients will never hear of protons at all.

It is remarkable that proton beam therapy as an accepted prostate cancer treatment modality is almost never among the treatment options described or recommended by surgeons, urologists, radiation oncologists, or medical oncologists. During the ten weeks that I attended the regular Wednesday night Education/Support Meetings, I can recall only three patients (of at least fifty) saying that their urologists recommended proton treatment, and one of those specialists who did so was a graduate of Loma Linda University. My own experience and that of Bob Marckini was the same, protons were not mentioned. The advocacy of those who have experienced the healing power of the proton beam can change this. This advocacy is led and championed by Bob Marckini, but those of us who have followed his lead have now taken up this challenge!

We, as members of The Brotherhood, have an opportunity to become advocates! I for one, intend to do this aggressively by writing my thoughts and experiences (this book), and telling anyone who will listen about the power of "The Beam." As "Brothers," we should all do as much as we can.

With the new proton beam therapy "Centers of Excellence" that will be coming on-line within the next ten years, this task will become easier, but as members of "The Brotherhood of the Balloon," it is our job to follow

Bob's lead, and continue to spread the word as "Proton Beam Advocates!" Just as an example, I recently stood up briefly at my monthly Genealogical Society meeting and simply told the seventy or so members present (about fifty ladies and twenty or so men) that I had been away for three months receiving proton beam therapy for prostate cancer, and that it was a totally non-invasive, minimum side effect treatment. I said that during our fifteen-minute break that I would be happy to give them my card and provide more information. During the break, I was swamped! Seven ladies and five men wanted more information! What was most interesting was that the featured speaker at this meeting, a former nurse, told me later that her father had received one of the first proton treatments over twenty years ago at the Harvard Cyclotron laboratory, the forerunner of the present Massachusetts General Hospital Proton Facility at Boston. At that time proton beam radiation therapy WAS experimental! No longer!

A previous reading of my paper, "A Status of Proton Facilities ..." received a similar response at a gathering of retired Space Program workers and engineers. One of these, a friend of mine for more than forty years, had been recently diagnosed, and was considering other treatment methods. I sent him information on proton beam therapy, and he recently completed treatment at the new Florida Proton Therapy Institute (University of Florida Shands), at Jacksonville, Florida. Since then, another long-time friend has gone there as well. I will be monitoring their treatments closely, so as to compare the methods used with those at LLUMC.

During the weekly support and education meetings at Loma Linda, time after time a "Newbie" would stand up and relate how he heard of protons by word of mouth from a friend, or even from someone who he did not even know, who had heard of his prostate cancer. I know this

The Brotherhood

works because recently I told a friend about my proton treatment. He later learned that an acquaintance had been diagnosed, and gave me his telephone number. I called him, and later he also decided to go to the Jacksonville proton facility. Another new Internet friend saw some of my comments on the Yahoo Prostate Cancer Support Forum, and after correspondence, has now started treatment at LLUMC's Slater Proton Center.

Even a business meeting is an opportunity! In a recent meeting with a representative at my bank with a young man of 44, as I was preparing to leave. I asked him: "When is the last time you had your PSA checked?" His answer: *"Three or four years ago, my present doctor does not believe in checking it."* This prompted a discussion that lasted another thirty minutes, and included protons!

Any gathering with relatives or friends is fertile ground for "spreading the word." We should certainly make sure that the members of our own family know of proton beam treatment! Remember that the presence of the disease in a family increases the likelihood of the appearance of prostate cancer in the male members. Make sure those family members that are in the forty to fifty-five age groups are getting their yearly physicals and PSA checks! Tell your story, so they will start to become aware of the dangers of prostate cancer, and the availability of The Beam!

When you have the opportunity, make a brief announcement in church. All that is necessary is a mention of proton beam therapy and the fact that the treatment has minimal side effects. If the congregation knew of your illness, this would be a good opportunity to thank them for their prayers. Also tell your church Illness Committee of your knowledge of protons as a potential cancer treatment, and that you are willing to share this knowledge. You may save some suffering and perhaps even save a life.

The following is a quote from Bob Marckini's December 2007 BOB Tales Newsletter:

"ONCOLOGY SURGEON CHOOSES PROTON THERAPY FOR PROSTATE CANCER"

"Dr. Ted Copeland is founder of the University of Florida Surgical Oncology Program and former Director of the UF Shands Cancer Center. When he was diagnosed with prostate cancer a few months ago, he knew his treatment options: surgery, X-ray radiation, seeds or proton therapy. According to an article in a University of Florida Alumni publication, Dr. Copeland wanted to make a choice that would give him the "best possible results." So, he chose to undergo 41 days of treatment at the UF Proton Therapy Institute in Jacksonville.

'Cancer patients need the very latest in information and treatment options when faced with this disease' says Copeland, the Edward R. Woodward Distinguished Professor of Surgery.

Hmmm... a renowned oncology surgeon chooses proton therapy when he's diagnosed with prostate cancer. What does this tell us?"

And so the Brotherhood continues to grow, and with it the continued expansion of the knowledge of proton beam therapy. As each new patient discovers the healing power of protons, another advocate is created. The existing five centers in the U. S. are already beginning to become inundated with requests for appointments, and only the opening of the new centers under construction and planned will provide relief. The treatment of prostate cancer with proton beam radiation therapy is certainly on the way to becoming the new "platinum standard." Our Brotherhood of the Balloon has contributed immensely to this, and we should be proud of this accomplishment.

Miscellaneous Thoughts

THIS BOOK MAY SEEM TO BE a testimonial to Loma Linda University Medical Center, and it is, to a degree. LLUMC is, after all, my "Alma Mater" since I am a "Prostate Proton Graduate." However, it actually is more of a testimonial to what I would call the "Power of The Proton Beam," no matter where the treatment is received.

The other Centers of Excellence that now provide proton beam radiation therapy may be in the early stages of developing their own variations of the treatment planning and procedures, but this is as it should be. The basic requirements have been established and approved, and all must adhere to them until improvements are established by proper clinical trials, approved, and implemented. The more experience that is brought to bear on the treatment itself—and the more that the treatment is used and made known—the sooner these improvements, which are inevitable, will occur.

Another miscellaneous thought: While I am most gratified that you, the reader, saw fit to read this book (and to see it through to the end), I must reiterate my recommendation that you also read Mr. Robert Marckini's great book *"You Can Beat Prostate Cancer, And You Don't Need Surgery To Do It."* Bob's work is a definitive description of an expert decision process, with all the background and technical details to help. In his book you will find a decision "Matrix" that does enable a realistic method of determining your own evaluation of treatment side effects. This matrix was developed by Nick DeWolf and was first presented on www.protons.com, Nick's prostate cancer site. Nick died in 2006, and this link now will connect you directly with the LLUMC Proton Site.

Bob Marckini's work is the result of not only his experience and his personal research, but also the interviews that he conducted both personally and through questionnaires with hundreds of prostate cancer patients.

Bob Marckini refers to himself as "A Recovering Engineer." For those who do not understand the implication, it simply means that "Once an engineer, always an engineer," no matter what you do in the meantime. As an engineer myself, I completely understand. Some doctors do also; I have had more than one medical specialist who, when they had been subjected to some of my questions about some treatment or procedure, would say: "You are an engineer, aren't you?!" I submit that in researching the possible treatments for your cancer, you must also utilize one of the primary tools that engineers are taught: *"Data Gathering."* Before any logical conclusion for a complex problem can be reached, one must know the relevant facts! I spent almost forty years in the business of preparing space vehicles for launch, including Atlas, Centaur, Titan, and the Space Shuttle, and in doing this one always had to consider the possibility of failure, and try to identify potential failure modes. Never was I more aware of this than when I was trying to decide on a treatment for my prostate cancer. This leads to the following, yet another in this series of "Miscellaneous thoughts."

Opinions: We have been advised many times that second opinions are good, and perhaps even medically necessary. The problem is that medical opinions may be biased toward the specialty of the doctor, and most do not mention proton therapy. If you have not had a second (or even third or fourth) opinion, and are still considering surgery (even robotic Da Vinci), or cryotherapy, or any other treatment alternative, consider this: A consultation and obtaining an opinion regarding proton beam radiation therapy for your treatment just might be a life changing experience, and for the better.

Miscellaneous Thoughts

The preceding subject was "Opinions," and how second and perhaps third opinions are good. But when urologists or other specialists do not count protons in their list of recommended treatments, as always, it is up to the individual to do his research and decide for himself. Here is a bit of anecdotal evidence regarding some of those urologists who **are** aware of proton therapy, but perhaps do **not** list it in their recommended list of possible treatments. This is from a prostate patient on the YANA site: *"I listened to the options my Urologist offered me for treatment, did my own research on those he mentioned and subsequently rejected his choices as not acceptable to me. During a follow up visit he* [the urologist] *incidentally revealed that his father had been diagnosed with Prostate Cancer some two years earlier. When my wife inquired as to the method of treatment the Doctor had chosen for his father, he replied* **'I sent him to Loma Linda University Medical Center for Proton Radiation.'"**

Until a few years after retirement, I enjoyed excellent health. Problems were minor and resolved quickly. A double hernia operation was the most serious situation that I faced, and except for a relatively minor stroke in 2003, I had no more health problems until 2006, when my PSA increased from 4.0 the previous year to 5.0, leading to biopsy and diagnosis of cancer. That, you see, is the problem with this prostate cancer disease: it is what I would term a "Stealth Invader." There is no early warning system! There are no serious symptoms until almost too late, and if you choose to ignore the rising PSA (or don't know that it is rising), it *WILL* be too late!

My urologist told me that if I did nothing, with a Gleason of 8, I would be dead in about two years! What is now plain is that I probably had the disease for several years before the diagnosis, and probably would NOT have been dead in two years. But then, after the initial shock, instead of accepting the first recommendation of cryosurgery, I chose to revert to the Marckini

"Recovering Engineer" mode, and to gather the data I needed to find the solution right for me. It was not an easy task, as many men have discovered.

The journey that you have started with prostate cancer is a long and arduous one, with perhaps many dark and narrow paths, with hidden obstacles and stops along the way. Once you survive the initial shock of diagnosis (and you will), you begin the really difficult task of determining the treatment "best for you."

An important part of any fact gathering exercise is proper record keeping. In this process of seeking information about prostate cancer treatments, it is very easy to get "lost" in the amount of information available. It is also of paramount importance that you have your medical reports available in an easily accessible location.

I suggest two notebooks (three ring binders work well) for your journey. The first should be a place to keep your notes about each treatment that you are considering, separated by dividers in the notebook, and arranged as you like it, but easy to refer to whatever you have collected, and to add new findings or references, and your own thoughts and conclusions. When you have questions or conclusions, write them down; it will help you as you go through this process.

The second notebook is the most important, because you will need to have these records at every stage of your journey. The records are the laboratory and doctor's reports for each and every PSA, CT Scan, MRI, biopsy, and any other doctor or hospital ordered test done to you that is related to your prostate cancer. Organize the notebook with dividers as you like, but keep each section in date order, so that when you are asked to supply a copy to a consulting specialist or facility, you can go immediately to the records, make the copies, and mail or fax them quickly. When you apply to a medical facility for a consultation or treatment, you will have to do this.

Miscellaneous Thoughts

I emphasize here: THESE ARE YOUR RECORDS! You may have to ask for a copy, but they are yours, and they cannot be denied to you. For instance, when the doctor does a biopsy, the analysis lab provides a complete report and you should receive a copy. Study it carefully, and if there are things that you do not understand, ask the doctor for explanations, then look them up for yourself. This is a vital part of your education and learning about the disease. Do this for any report that is related to your cancer.

In addition, the doctor himself prepares a written report of the procedure, which usually describes what was done, number of cores, how you "tolerated" the procedure and the size of your prostate gland as determined by the ultrasound equipment. Get a copy of this, and any written report of every visit that you make to the doctor. He/she should provide a copy to you when you ask, but may not do so unless you do!

The same is true for every test, scan, or other treatment that is done to you. In every instance, there *will* be a written report. Ask for it, and if not immediately available, ask that it be mailed to you. If you receive hormone shots (ADT) insist on a written record, signed by the doctor, of the type, size (one month, four month etc.), and when it was administered. The facility or hospital that provides follow on treatment will ask for these records. The same is true for any special tests such as ProstaScint, or MRI with Spectroscopy.

This may seem a bit overboard, especially considering the frustration and confusion that you are having after diagnosis, but I assure you these records will be invaluable on your journey. In addition, starting and organizing them will serve to help you make some sense out of the terms and things you may never have heard of before. It will help you organize your own thoughts and will assist in the many decisions that you are facing.

One might ask, "If proton therapy was not available, what treatment would you choose now?" This is the essential problem that I struggled with for the better part of three months after diagnosis. If there were no protons, I might choose brachytherapy, but with "stranded" seeds to prevent migration; or perhaps IG-IMRT radiation. The truth is, I just don't know, except that it would NOT be surgery. I would have to once again go through the soul-searching task of studying each option, checking for new advances, and weighing the all-important quality-of-life issues. But, thanks to protons, I do not have to do that!

There are a myriad of sources for information on prostate cancer, many of which have been mentioned. An excellent source is Aubrey Pilgrim. Aubrey was diagnosed in 1992, and his treatment choice then was RP, "a choice that he regrets to this day." [63] He has provided a comprehensive review of almost all aspects in the fight we face. This is a book available on-line and completely free for downloading, that he calls: "A Revolutionary Approach To Prostate Cancer." [64] I highly recommend that you read this work. It was written in collaboration with medical doctors, and is 380 pages of concentrated information. You can pick and choose chapters based on the topics of most interest to you. Aubrey's compelling personal story is also available on the YANA Mentor's Experiences Web site. Aubrey, whom I considered a personal friend, and who helped me in the initial drafts of my book, died of a brain tumor in 2008. He is missed.

Other Internet forums are available and useful in searching for the answer "best for you." A word of caution: Many of these forums or groups offer help and support in dealing the shock and fear of a new diagnosis, but are primarily "hand-holding" forums. Some of the members will offer advice, recommend, and profess knowledge of the systems of treatment that worked for them. In reading through past postings, you will find many that describe various problems such as erectile dysfunction, incontinence, diarrhea, and other such issues that proton beam therapy minimizes. Be cautious in accepting at face value recommendations from the forums; do your research, and make up your own mind.

What is most disturbing are the posts of some of the men that have neglected the yearly PSA checks; or have not "done their homework;" or perhaps did not have an experienced, expert physician for their treatment of choice. There are the instances where the person suddenly finds that he has advanced PCa; or has never recovered continence and is dealing with diapers and all that entails: *"No, I would not have surgery again if I was to do it over. 3 1/2 years in diapers is not what I ever expected;"* or has come to the realization that ED is a way of life for him now because of the treatment he chose; or has completed treatment some time ago and the cancer is recurring.

These instances are far too prevalent for my own peace of mind. Usually a review of a day or so of posts to a cancer support forum, or any visit to a local support group, will provide several instances of these types of problems. Quotes such as: *"[my husband] was terminal by the time he was diagnosed with Prostate cancer - his PSA was 412 with extensive mets to the pelvic area & spine."* are heartbreaking. I live in hope that medicine and cancer research will soon take a "quantum jump" and put such things in the past. To me, proton beam therapy appears to be a huge step in the right direction.

In 2007, I joined one of these forums, the Prostate Cancer Support Group, and offered information on proton therapy. I found myself "defending" protons against the various other modalities. It took five posts to get one individual to admit that protons might be a viable alternative, and this only after I posted links to each of the five hospitals that are now offering PBRT. Even then, the question was asked, *"If it has been out there for years, is effective, and is being utilized by world-class institutions such as M. D. Anderson, Mass. General Hospital, etc., then why would it not be recommended as a treatment modality?*" The answer, regretfully, boils down to one thing: money. Medicine is a business, and hospitals and doctors "sell what they have." Unfortunately this is not an answer that goes over well.

This thought leads to the recent outcry by certain institutions and individuals for "Randomized Clinical Trials" (RCTs) for proton therapy in comparison to the various other radiation treatments. The best answer I have found to these arguments was published in January 2008 by Michael Goitein of the Department of Radiation Oncology, Harvard Medical School, Boston, and James D. Cox, of the Division of Radiation Oncology, University of Texas M. D. Anderson Cancer Center, in Houston. This paper was published in *Journal of Clinical Oncology*, Vol. 26, No 2 (January 10), 2008: pp. 175-176.[65] The authors refute the argument for RCTs by pointing out that such trials would be unethical if the participants were aware of all the facts pertaining to the trials, and also that the proponents of RCTs are primarily concerned with the comparative costs of the treatments.

Discussions on these group listings can be very informative when the informer is an expert in the subject at hand, either with a particular form of treatment, or very experienced and knowledgeable after years of battling the disease. Spending enough time following such a list will soon make you aware of which poster one should pay attention to, and which of them one should not. Some men have really dedicated themselves to helping others in anyway that they can. Two of these have already been mentioned in this book, Terry Herbert and Aubrey Pilgrim. Another is Charles "Chuck" Maack, who has made an exhaustive study of advanced prostate cancer, androgen deprivation therapy, and many other aspects of the disease. To learn more about any of these men, simply Google his name in combination with "prostate."

Miscellaneous Thoughts

New ideas or information concerning new forms of treatment may also be found on these forums. In this way I learned of a newly developed machine using a combination of CT Scanning and IMRT, called the "TomoTherapy® Hi·Art®" system. Combining advanced IMRT with CT technology, it seems like a very good system, but in discussing the machine with a member of the list, I learned that the developers have incorporated hypo-fractionation (larger daily dose fractions) into the treatment protocol. This is as yet experimental, and being new, no long-term results are available. As pointed out in the chapter on protons, any treatment that increases radiation dosage to non-cancerous tissue may possibly be harmful either in the short or long term.

I recently had occasion to speak with a gentleman who was in the final phase of trying to make up his mind on a treatment, after having waited over three months for "hospital clearance" of a new method of treatment, "irreversible electroporation." [66] The FDA has apparently recently approved the procedure, but I could find no instance of anyone yet being treated with it for prostate cancer. With no history of results, I would be reluctant to be the first patient to be treated with any new method, no matter how promising, but that is up to the individual. I have great respect for the courage of those who partake in clinical trials to prove these new treatments.

There are other treatments that are not widely used and some that have not yet been granted FDA approval in the U. S. High Intensity Focal Ultrasound, HIFU, is one of these. While it has been used in Europe and elsewhere for a number of years, my research seemed to indicate results that were somewhat varied (perhaps due to the "expertise" factor).[67] There does seem to be more likelihood of impotence and perhaps incontinence than with proton therapy. In my own opinion, the same is true for cryotherapy, even though it is an FDA approved treatment modality.

Chapter Twelve

Once a person (myself included) makes up his mind that a particular treatment is for him, and has the procedure—whatever it may be—he seems automatically to feel that he must justify that decision. This can be true even if the side effects associated with the treatment are causing problems. This is plain and prevalent on the forum mentioned above. Each of us becomes more or less of an advocate for whatever modality we have chosen, and this is quite understandable. But there are cases, when a treatment has either failed or caused terrible long-term side effects, when a man may state that he would not make the same choice again.

Another caution: In some instances, medical facilities or companies actually have engaged in "misinformation" using the Internet to promote the particular product or modality that they offer, attempting to explain why their product or system is "better" than other products or treatments such as surgery or protons. Some statements to be careful of begin with *"We are not aware of ..."* and *"We believe that...."* With careful reading and investigation, you will find that they do not tell the whole story, or just are plain incorrect. The point is that not all information found on the Internet is "fair and balanced." Make sure that you investigate and understand the Web site's or individual poster's motivation, and follow up any referenced data to see if it is from a valid source.

There are also "Support Forums" for our wives, partners, and "better half" female companions who also are on this journey with us, and must also traverse it every step of the way. Here is a "Ladies Only" Support Group that our female partners may join to learn and receive help, and the knowledge and experience of others like them: *http://www.forums.prostate-help.org/.*

Another miscellaneous thought: If you are someone in your family has been diagnosed with cancer, please be aware that the children in the family need to be helped to understand the situation. The subject of cancer is very frightening, and even very young children will pick up on the fearful undertones of conversations and the attitude of their parents. They need special reassurances, and need to know that the parent is not excluding them from the knowledge of what is going on. From the ACS site:

"If children hear about their parent's cancer from someone else, such as a curious neighbor or a classmate, it can destroy the trust that parents have worked so hard to put in place. If children think their parents are being vague on purpose or are trying to hide something from them, they find it hard to believe they are being told the truth. So it is better that parents learn how to share this information truthfully, but in a way that allows the child to understand and take part in the discussion. The other problem in keeping the cancer a secret is that the child may incorrectly assume that whatever is happening is too terrible to be discussed. This may lead the child to feel isolated from the family, so the natural desire parents have to protect their children sometimes only makes things harder....." [68]

Thus, as one has to study the cancer subject in order to learn how best to treat it, one must also make an effort to learn how best to inform the children in the family. There are different ways to handle the situation for each age group of children, and indeed for each individual child, depending on his or her needs and capabilities.

If you have read this far, and are still not sure if proton beam radiation therapy is for you, why not schedule a consultation with a radiation oncologist at one of the Proton Centers nearer, perhaps, to your home. *If you do not, you have not accessed all the facts necessary for a fully informed decision.* If you do have this consultation, *you will learn about protons from a professional.* Make a list of your questions so that you will not forget any important points.

You will likely be impressed with the entire facility and the staff, and when you learn about protons from the experts, *you may find that the path along your personal journey has suddenly become much clearer.*

If you do choose Loma Linda University Medical Center's Department of Radiation Medicine for this very important "second opinion," I *know* that you will be impressed with the hospital, the medical staff, and the caring attitude that is part of the overall philosophy of this institution. The doctors there, after all, have been successfully treating cancer victims with protons since the fall of 1990! And their motto is "To Make Man Whole!"

Another of the many benefits of LLUMC is the Drayson Center, where almost any form of exercise is available. The importance of exercise for proton patients was stressed at Loma Linda, but even if you choose another treatment center, exercise as often as you can. The fatigue that you may begin to feel about halfway through the proton treatments can be offset by regular exercise; three times a week, for an hour or so will help tremendously. If you go to some other Proton Center, keep this very important exercise requirement in mind.

The regular Wednesday Support/Education Meetings at Loma Linda were great! A different Proton Center may have similar meetings (at Jacksonville, they are breakfasts or luncheons). Just *attend* these when you can. Sometimes you will hear a testimonial that goes right to your heart, or one that is truly amazing! The fellowship with other patients is always outstanding. At our weekly meetings, Dr. Lynn Martell always had an interesting program, and several pertinent stories. And Patti Lee kept everything organized, from "Lunch Bunch Tours" to Golf and Tennis Matches. The educational part of these meetings was also great. We had presentations by the Loma Linda Historian, Mr. Richard Schaefer, whose work is summarized in this book; the Chief Medical Physicist in the Department of Radiation Medicine, LLUMC, Dr. Baldev Patyal; the Chief Accelerator Physicist, Dr. George Coutrakon; and others.

And finally, there is the importance of laughter. Here is what one medical doctor has to say about laughter and cancer, with which he himself was afflicted.

"Laugh at it. My work with cancer patients taught me the therapeutic importance of laughter. A good giggle makes patients feel better, not only emotionally but also physically. It temporarily makes their pains, even severe cancer pains, disappear. From a purely physiological standpoint laughter creates increased relaxation and oxygenation. Endorphins, the body's homegrown "narcotics," go to work. The body's immune system is stimulated as well. Laughter brings about well-being by combating destructive stress, depression, rage, and insomnia. It provides an overall liberating effect. Distraction from oneself, from one's physical and other concerns, plays a beneficial role too. I laughed a lot at home, in my office, as well as at the hospital. I woke up every day to affectionate messages and hilarious e-mails. Then it was off to the hospital for radiation therapy or for treatment of its rather severe side effects. (At one point radiation therapy made me hardly able to walk and unable to sit)" [69]

Another quote: *"Laughter in and of itself cannot cure cancer nor prevent cancer, but laughter as part of the full range of positive emotions including hope, love, faith, strong will to live, determination and purpose, can be a significant ... aspect of the total fight for recovery."* [70]

Thus, "laughter is good medicine," and the LLUMC Wednesday night "Support" meetings had plenty! Dr. Martell saw to that (along with the other "jokesters" always in the crowd). The fun and fellowship that just "happened" was amazing! Beneficial exchanges of information were the norm, and in addition, lifetime friendships were formed. These meetings were, in and of themselves, an education. Yes, at times they were somewhat repetitive in the ways that the individuals "found" the proton beam; the thankful graduation "speeches" given by those that were completing their nine weeks of receiving the curing treatment; and the glowing reports from "alumni" that completed their treatments years before; but these things speak volumes about both proton beam therapy and LLUMC.

The "miscellaneous" thought once again strikes me that, as a proton prostate cancer patient and survivor, I am one who is a member of a very unique group of people that through investigation, research, recommendation of a physician or a friend, or some other circumstance "discovered" the proton beam. Not many do.

But Dr. Martell's favorite saying is *"Nothing Happens By Accident!"* With that in mind, I decided to become a "Proton Beam Advocate," and do my best to make others aware that there is another alternative that should be considered. This book is one result.

Looking back on my time at Loma Linda, I can truthfully say that it *WAS* a life-changing experience. I arrived there full of apprehension, not really knowing what to expect, but with a feeling of hope that I really had no basis for, except from my own research and the testimonials of those men that had traversed this path before me. I know now that they were correct and factual in their stories about the Loma Linda experience. I join them in saying thanks to Loma Linda University Medical Center, to the administration and all the staff, and most especially to Dr. Rossi, Dr. Jabola and Nurse Sharon Hoyle, to Lynn Martell and Patti Lee for all their fine efforts, and "The HBL Crew" (they know who they are). I left Loma Linda with a mixture of sadness and hope. I was sad to leave what had become an enriching experience, but I was filled with new hope for the future because I knew that I had received the best treatment for my cancer at what I consider to be one of the finest hospitals in the world.

Afterword

IN THIS BOOK I have attempted to document my efforts to find a treatment for prostate cancer, and to explain why I believe that proton beam therapy should be considered by anyone diagnosed with the disease.

Prostate cancer is such an insidious disease that many men may not become aware of the danger until they are suddenly diagnosed. I sincerely hope that my work will help men become aware of the need to make the PSA test and DRE a mandatory part of a yearly physical exam.

I also hope that this book will provide informational help for those faced with the diagnosis of cancer. Anyone diagnosed needs all the help that they can get. Instead, what they usually find upon diagnosis is a morass of conflicting and confusing opinions, from both medical specialists and other patients as well. To help with this most distressing situation, I want to re-emphasize some factors from the early chapters of this book.

- If diagnosed with prostate cancer, do not panic. Prostate cancer should NOT be considered to be a death sentence.

- Do not decide immediately on a course of treatment just because the diagnosing medical specialist recommends it. You should seek "second opinions" from specialists that offer different treatment methods.

- Take your time and learn all that you can about all treatment options available to you. You will find several options that will likely provide a very good chance of arresting and/or eliminating the cancer.

- Rest assured that the outcome, with experienced medical specialists and modern equipment, is approximately the same for most treatment modalities. The major differences in results may be in "Quality of Life" issues, which may or may not happen in every case.

- Whatever method you choose for your treatment, you should be comfortable with it. Be glad that you made the choice, based on your personal situation. This is many times better than accepting the recommendation of the diagnosing specialist, then later wondering why you did not do your "due diligence" and make your own decision.

- Regardless of which treatment you choose, recognize that you are on this journey for life. The periodic PSA checks, and perhaps other new tests, are mandatory and must not be neglected. Research continues to provide increasing knowledge in the battle against cancer. New or greatly improved medical equipment, techniques, tests, and drugs continue to be developed. Keep up to date.

- There is every reason for you to put "Quality of Life," both during and following treatment, at the top of a list of criteria to be considered in reaching a decision. When you do this, proton beam therapy will invariably rank high on the list of choices. However, you should study each option, and—after consultation with the appropriate medical specialists and evaluation of their recommendations—make your own decision based on the characteristics of your disease, your personal and family situation, and general health.

End Notes and Citations

[1] Us Too, A Prostate Cancer Support Group – Web site.
<http://www.ustoo.com/Overview_Statistics.asp> Acc. Dec. 2006.

[2] John Shuey. Internet post on Yahoo Prostate Cancer Forum.
http://health.groups.yahoo.com/group/ProstateCancerSupport/message/13966
Posted July 22, 2007; used with permission.

[3] Robert Young. Phoenix5. A Web site; accessed August 2007.
http://www.phoenix5.org/battle.html [Permission requested, but Webmaster deceased.]

[4] Dr. Jonathan Oppenheimer, MD. The Prostate Lab. A Web site.
< http://www.prostatelab.com/grading.htm> Accessed August 2007.

[5] Oncura. Prostate: Cryotherapy. Web site; accessed Dec. 2006.
<http://www.oncura.com/prostate-cryotherapy.html>

[6] Dr. Marc. Greenstein. "Who Should Have Robotic ..." At
<http://www.healthcentral.com/prostate/c/122/14777/surgery-cancer/1/>.

[7] Jeffrey S. Eshleman, et al. Radioactive Seed Migration to the Chest....
J. Radiation Oncology Biol. Phys., Vol. 59, No. 2, pp. 419425, 2004
http://www.prostate-cancer.com/brachytherapy/survival-rates/survival-rates-seed-migration.html Accessed August 2007.

[8] Seed Migration in Prostate Brachytherapy. Article in the British
Journal of Radiology (2003) 76, 913-915. Accessed August 2007.
< http://bjr.birjournals.org/cgi/content/full/76/912/913>

[9] Prostate Cancer Foundation; a Web site.
<http://www.prostatecancerfoundation.org/site/c.itIWK2OSG/b.47300/k.9D4F/Radiation_Therapy.htm> Accessed October 2006.

[10] Gleason Experts. Don Cooley's Web Site; accessed June 2007.
<http://www.cancer.prostate-help.org/cagleex.htm>

[11] San Diego CyberKnife Center Web site. Accessed August 2007.
<http://sdcyberknife.com/dr-cancers/prostate2/index.htm>

[12] Terry Herbert. YANA – You Are Not Alone Prostate Cancer
Support Site. <http://www.yananow.net/Mentors/TerryH.htm>

[13] Dr. Patrick Walsh. Interview; PBS Online News Hour.
http://www.pbs.org/newshour/health/prostate/walsh_extended.html
Accessed May, 2007.

[14] American Cancer Society Web site. Search for Hormone Therapy
on: <www.acs.org>.

[15] Wikipedia, The Free Encyclopedia. A Web site.
< http://en.wikipedia.org/wiki/Brachytherapy> Accessed Aug. 2007.

[16] Dr. Adam P. Dicker. Medical College of Thomas Jefferson
University, Philadelphia, PA. Posted on "SeedPods" Sept. 1998.

[17] Richard E. Peschel et al. "Iodine 125 Versus Palladium 103
Implants for Prostate Cancer..." The Cancer Journal, Volume 10,
Number 3, 1 May 2004 , pp. 170-174. Accessed August 2007.

[18] Peoxcelan Cesium −131. A Web site. Accessed Sept. 2007. <http://www.cesium131.com/>

[19] Eshleman. Ibid.

[20] E. M. Tapen, ; J. C. Blasko, et al. "Reduction of radioactive seed embolization to the lung …" Int-J-Radiat-Oncol-Biol-Phys. 1998 Dec 1; 42(5): 1063-7. Accessed online August 2007.

[21] J. Coen et al. "Comparison of Proton vs. brachytherapy…" Abstracts > 2006 Prostate Cancer Symposium. http://www.asco.org , Accessed August 2007.

[22] Subatomic Particles; a Web Site. Accessed June 2007. <http://www.sciencebyjones.com/subatomic_particles.htm>.

[23] Life Extension; a Web Site. *Principals of Radiation Therapy.* http://www.lef.org/protocols/cancer/radiation_therapy_01.htm. Accessed June 2007.

[24] Ulrick Mock et al. Department of Radiotherapy and Radiobiology, Medical University of Vienna, Vienna, Austria. Abstract: Comparative treatment planning…photon- versus proton-based radiotherapy. PMID: 15995838. Accessed August 2007.

[25] Proton Treatment for Prostate Cancer. A Web site. <http://www.prostateproton.com/> Accessed October 2006.

[26] Prostate Cancer and patients who have chosen proton treatment. <http://www.protonbob.com/proton-therapy-aboutus.asp>. Accessed November 2006.

[27] The Cancer Journal. March/April 2007 "Is It Time to Use Protons for Breast Cancer?" Commentary. Accessed May 2007. <http://www.journalppo.com/pt/re/ppo/abstract....>

[28] Wikipedia, The Free Encyclopedia. A Web site. <http://en.wikipedia.org/wiki/Magnetic_Resonance_Imaging> Accessed December 2006.

[29] WikipediA, The Free Encyclopedia. Web site. Acc. Dec. 2006. <http://en.wikipedia.org/wiki/Computed_tomography>

[30] Dr. Carl J. Rossi. Article: "Conformal Proton Beam Radiotherapy of Prostate Cancer." Originally found on the Internet; apparently no longer available, although excerpts were shown in a PDF document, accessed April 2007, written by Aubrey Pilgrim: <www.**cancer**.prostate-help.org/download/**pilgrim**/11proton.pdf>

[31] The Free dictionary (On-Line). Accessed January, 2007. <http://medical-dictionary.thefreedictionary.com/Proton+(physics)>

End Notes and Citations

[32] Oncolink. A Web site. Abramson Cancer Center of the University of Pennsylvania. Article by Dr. James Metz: "Differences between Protons and X-rays." Accessed March, 2007.
<http://www.oncolink.org/treatment/article.cfm?c=9&s=70&id=210>

[33] Roy Butler. Chemistry Department; Norwich University, Northfield, Vermont. From: "The Patient Proton; *"Various aspects of proton treatment ... ""* Figure 8.

[34] The National Association for Proton Therapy, a Web site.
<http://www.proton-therapy.org/prostate.htm>; Accessed Mar 2007

[35] Eugen B. Hug. Department of Radiation Oncology, Dartmouth Hitchcock Medical Center, Lebanon, NH USA. Abstract: PMID: 15971306 PubMed; a Web site. Accessed March 2007.
<http://www.ncbi.nlm.nih.gov/entrez/query.fcgi?cmd=Retrieve&db=PubMed&list_uids=15971306&dopt=Abstract>.

[36] Dr. Carl J. Rossi. Department of Proton Radiation Loma Linda University Medical Center. As provided in the Prostate Cancer Communication Newsletter; Vol. 23, No. 1. March 2007.
See: http://www.proton-therapy.org/; scroll to Dr. Carl Rossi.

[37] J. F. Torres-Roca. Moffitt Cancer Center, Tampa FL USA. Abstract: "The Role of EBRT ...". PubMed Article, July 2006.
http://www.ncbi.nlm.nih.gov/entrez/query.fcgi?db=pubmed&cmd=Retrieve&dopt=AbstractPlus&list_uids=16885914&query_hl=5&itool=pubmed_docsum; PMID: 16885914. Accessed May 2007.

[38] CERN Press Release Web site. Accessed April 2007.
<http://press.web.cern.ch/Press/PressReleases/Releases2006/PR15.06E.html>

[39] Dr. Baldev Patyal, Chief Medical Physicist, Dept. of Radiation Medicine, LLUMC. Presentation; February 2007.

[40] Protons, A Beam of Hope. A Web site of LLUMC.
<http://www.llu.edu/info/legacy/>. Accessed Feb. 2007.

[41] National Association for Proton Therapy; a Web site article;
<http://www.proton-therapy.org/howit.htm> Accessed Oct. 2006.

[42] Oncolink. A Web site. Abramson Cancer Center of the University of Pennsylvania. Article, Cancer Treatment Information.
<http://www.oncolink.org/treatment/article.cfm?c=9&s=70&id=209>
Accessed March 2008.

[43] Gy: Gray: a unit of absorbed dose of ionizing radiation equal to an energy of one joule per kilogram of irradiated material. Abbrev. *Gy.*

[44] Army Health Care. A Web site;
<https://www.goarmy.com/amedd/news.jsp> accessed March 2007.

End Notes and Citations

[45] University of Pennsylvania Health System. A Web site; <http://www.pennhealth.com/cam/proton/index.html>

[46] KFOR-TV; Oklahoma City, OK. AP News Release. See <http://www.msnbc.msn.com/id/18040175/>Accessed Aug 2006.

[47] Business Wire, April 9, 2007. Procure Begins Construction on ... First Proton Therapy Cancer Center. <http://findarticles.com/p/articles/mi_m0EIN/is_2007_April_9/ai_n1 8792572> Accessed March 2008.

[48] News Release. Northern Illinois University. <www.niu.edu/PubAffairs/RELEASES/2006/dec/protonrfp.shtml>

[49] Oncolink; Ibid.

[50] Dr. James Slater [Loma Linda Univ. Medical Center]. Article Health: "More U.S. Hospitals Offering Proton Beam Therapy;" News USA. A Web site. Accessed October 2006. <http://www.proton-therapy.org/UShospitals_191A.pdf>

[51] MIT New Office' a Web site article. "MIT Proton Treatment ..."; August, 2006. Accessed. Feb. 2007.<http://web.mit.edu/newsoffice/2006/proton.html>.

[52] Synthesis, A Publication of the UC Davis Cancer Center. <http://www.ucdmc.ucdavis.edu/Synthesis/issues/fall_winter_06-07/features/in_translation.html> Accessed August 2007.

[53] Advanced Cancer Therapy. A Web site. Accessed August 2007. http://www.advanced-cancer-therapy.org/

[54] Richard Schaefer. *Legacy, Daring to Care*.; p 69. Legacy Publishing Association, Loma Linda CA; 2005. Note: Some citations from Mr. Schaefer's work are provided in the body of the text to avoid many "ibid" citations here.

[55] George Washington: Health and Medical History; A Web site. <http://www.doctorzebra.com/prez/g01.htm>. Accessed Jan. 2007.

[56] Richard Schaefer. Ibid; p 73.

[57] Ibid. p. 74.

[58] National Association for Proton Therapy. A Web site; < http://www.proton-therapy.org/pr02.htm> Accessed Sept. 2007.

[59] Loma Linda University Children's Hospital; A Web site; <http://www.llu.edu/lluch/> Accessed February 2007.

[60] Ibid.

[61] ProtonBOB.com, a Web site. Accessed Feb. 2007. <http://www.protonbob.com/proton-therapy-aboutus.asp>.

End Notes and Citations

[62] Robert J. Marckini, *You Can Beat Prostate Cancer And You Don't Need Surgery To Do It.* (Self Published), 2005, p. 148

[63] Aubrey Pilgrim. YANA – You Are Not Alone Prostate Cancer Support Site. Mentor's Experiences. Accessed May 2007. .<http://www.yananow.net/Mentors/AubreyP.htm>

[64] Aubrey Pilgrim. "A Revolutionary Approach To Prostate Cancer." An "E-Book" available free. Accessed May 2007. <http://www.cancer.prostate-help.org/capilgr.htm>

[65] Michael Goitein and James Cox. "Randomized Clinical Trials ...;" *Journal of Clinical Oncology*, Vol 26, No 2 (January 10), 2008: pp. 175-176. <http://jco.ascopubs.org/cgi/content/full/26/2/175> Accessed April 2008.

[66] Onik G, Mikus P, Rubinsky B. "Irreversible electroporation: implications for prostate ablation." [Journal Article] *Technol Cancer Res Treat 2007 Aug; 6(4) :295-300.* Accessed April 2008. http://www.unboundmedicine.com/medline/ebm/record/17668936/...

[67] Clinical Results of HIFU. Maple Leaf HIFU Co. Acc. August 2007 <http://www.hifu.ca/physician/clinical_results.php>

[68] Helping Children When A Family Member Has Cancer: Dealing With Diagnosis. American Cancer Society; Article on the Internet. http://www.cancer.org/docroot/CRI/content/CRI_2_6X_Dealing_With_Diagnosis.asp?sitearea=CRI . Accessed December 2007.

[69] The joy of cancer: .... A Web site Article. <http://www.thefreelibrary.com/> Accessed April 2007.

[70] Harold. H. Benjamin, PhD. Web site quote. Accessed Apr. 2007. <www.healingcancernaturally.com/laughter-is-medicine.html>

161

www.ingramcontent.com/pod-product-compliance
Lightning Source LLC
Chambersburg PA
CBHW020205200326
41521CB00005BA/247